LATERITE

Genesis, Location, Use

Monographs in Geoscience

General Editor: Rhodes W. Fairbridge

Department of Geology, Columbia University, New York City

B. B. Zvyagin
Electron-Diffraction Analysis of Clay Mineral Structures—1967

E. I. Parkhomenko
Electrical Properties of Rocks—1967

L. M. Lebedev
Metacolloids in Endogenic Deposits—1967

A. I. Perel'man
The Geochemistry of Epigenesis—1967

S. J. Lefond
Handbook of World Salt Resources—1969

A. D. Danilov
Chemistry of the Ionosphere—1970

G. S. Gorshkov
Volcanism and the Upper Mantle: Investigations in the Kurile Island Arc—1970

E. L. Krinitzsky
Radiography in the Earth Sciences and Soil Mechanics—1970

B. Persons
Laterite—Genesis, Location, Use—1970

D. Carroll
Rock Weathering—1970

In preparation:

A. S. Povarennykh
Crystal Chemical Classification of Minerals

E. I. Parkhomenko
Electrification Phenomena in Rocks

R. E. Wainerdi and E. A. Uken
Modern Methods of Geochemical Analysis

LATERITE

Genesis, Location, Use

Benjamin S. Persons
Partner, Dames & Moore, Consulting Engineers
Atlanta, Georgia

With a Foreword by H. W. Straley III
Professor of Geology
Georgia Institute of Technology

SPRINGER SCIENCE+BUSINESS MEDIA, LLC 1970

Library of Congress Catalog Card Number 73-107541

ISBN 978-1-4684-7217-2 ISBN 978-1-4684-7215-8 (eBook)
DOI 10.1007/978-1-4684-7215-8

Originally published by Plenum Press, New York in 1970
Softcover reprint of the hardcover 1st edition 1970

A Division of Plenum Publishing Corporation
227 West 17th Street, New York, N.Y. 10011

United Kingdom edition published by Plenum Press, London
A Division of Plenum Publishing Company, Ltd.
Donington House, 30 Norfolk Street, London W.C.2, England

For Frances, who, for years, lived beside a rain forest road

FOREWORD

In comparison with engineering, geology is a relatively new domain of knowledge. Man has been building almost from the moment he came down out of the trees or emerged from the caves. All of his structures were founded in or upon rock or soil. Before the end of what we call ancient history, he had learned a great deal about materials, mechanics, and structures. This empirical information had become an organized field of practical knowledge by the time of Leonardo da Vinci. Although both foundations and materials of construction were largely earthy, codified knowledge of neither one nor the other existed at that time.

During the last two centuries, geology has emerged and has recently begun to take on a more quantitative aspect. A generation ago, it joined hands with civil engineering to create soil mechanics. Engineers began to apply the science of geology to foundations and materials with astoundingly successful results, and geologists began to acquire an understanding of engineering methods, applying what they could to their problems. Geological engineering was born of this union.

People of an older time employed stone and brick in construction, although cut brick and sawed stone were used more sparingly because of a scarcity in both suitable raw materials and techniques. They were used in Cambodia, for example. A material able to meet requirements was found nearby, known as *itica culla*. In India it was called *vettu culla*, but F. A. Buchanan, a British engineer who discovered it there in 1800, gave it the latinized name of *laterite*.

Wide seasonal fluctuation of rainfall accompanied by high air temperatures seem to be controlling factors in laterite formation, a climate found in lands near the equator, where per capita consumption of extraneous energy is relatively low. It was in some of these tropical or near-tropical lands that Ben Persons was introduced to it as an engineering material, although he had as a lad, hunted birds and ridden horses over a remarkably similar substance. Mr. Persons was struck with the material and with the possibilities that it might present. With characteristic energy, he immediately set about making himself a laterite specialist. He has now seen, observed, and worked with lateritic materials of one kind or another in many parts of the world.

As might have been predicted from his character, education, and experience, he has evinced less interest in bauxitic, itabiritic, and/or manganiferous varieties. These are the laterite-like materials of greatest interest to mining, as opposed to civil engineering, and it is in this connection that I entered the picture, Mr. Persons calling me into consultation with reference to the geological aspects of laterites. In most instances, except where bauxite or itabirite was closely related, it was readily apparent that Mr. Persons' knowledge of the subject was more than ample to meet his needs.

His reading has been extensive, his observation keen and over parts of three continents. His conclusions are well and carefully thought out. Both his interests and his writing are those of an engineer. He has, nevertheless, made himself a specialist in the geological aspects of this unique material and his work has the earmark of authority.

In the present volume, he has covered the subject in both breadth and depth. He points out, by both description and maps, the hitherto unrecognized, widespread occurrence of laterite materials. He discusses origin in terms that engineer, geologist, and layman can understand. He details the uses to which an engineer may put laterite and the limits of its utility. He cautions agriculturists as to some of the pitfalls lying ahead for extensive and/or intensive farming under climatic conditions conducive to its formation.

H. W. Straley, III

PREFACE

This volume is intended to be a useful review of the subject of laterite. Though some confusion and marked disagreement exists among students and users of laterite, there is some common ground of agreement on what constitutes laterite, how it came to be, where it is found, and how man may use it for his purposes.

All too often books or papers dealing with laterite have taken either an academic or practical approach. In doing this, the authors have excluded a large audience interested in laterite as an environmental phenomenon or a useful material. Yet both the academician and the user should understand each other if they are to communicate and achieve their ultimate purpose.

Following an engineering education and nearly a score of years experience in earth design and construction in regions supposedly devoid of laterite, I was confronted with the task of the earth design of a road in West Africa, where laterites and lateritic soils constitute the principal and most readily available source of road building materials. This responsibility instituted a need to understand these materials, relate their properties to other known materials, and communicate this knowledge to those who would build and maintain the road. While trying to accomplish all this, I discovered that communication is all too often blocked by semantics and prejudice, semantics here relating to two people calling two different things by the same name and prejudice relating to having a preconceived notion about a thing and being unwilling to accept external knowledge.

Perhaps because laterite often occurs in remote and relatively inaccessible regions, knowledge of its characteristics, locations, and uses has, until recently, been communicated only between those who work in this same area. It is found in hot, wet, forested areas colonized by the great European powers, in not-so-wet grasslands, and in deserts which once were wet. Geological and engineering civil servants charged with knowing and using laterite often spent their careers as regional specialists. Seldom were their successes and failures made known to their professional brothers working with the same materials in areas separated from them by half a world.

Scientific and engineering knowledge acquired in the colonies was usually a closely guarded secret of the empire and, even when free communication was attempted, differences in language and terminology inhibited exchange of experience. When vast areas of the world were quitted by the colonial powers after World War II and local self-government grew, large foreign aid programs were aimed at raising the economic standards of these former colonies. Communication within and between newly independent countries was a first order of business, and simple, unsophisticated roads and railroads filled the need.

Engineers strove to build with the unknown but inexpensive materials at hand and geologists searched for materials the engineers told them were needed. An era of failures followed, the new roads proving to be more of a curse than a blessing. After much loss of face, those who sought to aid pulled back and decided they should understand the materials with which they were trying to work before building more spectres to haunt them.

Within the last decade, notable and useful work has been done toward seeking an understanding of laterite. A paper by Mary McNeil published in Scientific American* and the *Review of Research on Laterite* by UNESCO aided in defining, locating, and understanding the forming processes. Neither of these studies, nor the many other similar works, were intended to serve the user. Independently, engineers presented papers describing successful techniques and even set standards for identifying usable materials; yet, a gap remains between locating usable materials and knowing that which it is necessary to locate.

It is my belief that an accumulation of knowledge to date would be useful to the combined team of geologist, soil scientist, and engineer, and I have attempted to write, in a manner hopefully not an affront to the knowledge of any, an inclusive book for their work *together*.

As man progresses in using laterite successfully, and as its use grows more common, the need for such a book will diminish. In the foreseeable future, the engineer will write specifications which the geologist may easily follow. Future professionals will wonder why all the fuss. But, in the meantime, there are roads to be built and laterites which need to be found, and people to use the roads, none of which can wait.

Benjamin S. Persons

Atlanta, Georgia, 1969

*"Lateritic Soils," *Scientific American*, November 1964, p. 97.

Well-indurated, uniform laterite from the rain forest of West Cameroon along the Kumba–Mamfe road.

ACKNOWLEDGMENTS

Dr. H. W. Straley, III, of Georgia Tech aroused the initial interest in this book and agreed to furnish the foreword. He also furnished a most thorough review of the manuscript.

Members of the staff of the Soil Conservation Service, USDA, Athens, Georgia, and particularly Mr. Frank T. Ritchie, Soil Scientist, were helpful in reviewing the manuscript and providing worthwhile comment.

At the request of the author's firm, Professor Ta Liang of Cornell reviewed the manuscript. His comments, suggested revisions, and suggested areas of further study by the author as well as his encouragement were most appreciated.

Alex W. Bealer, III, of Atlanta, a successful author, led the writer through the labyrinth of getting a book published.

CONTENTS

Chapter 1

GENESIS

GENERAL

In 1800, while on a journey through Malabar and Kanara in present day India, Buchanan[1] reported on a material which could be cut "with a trowel or a large knife" and when exposed to air became hard. He called this an "indurated clay" The material, which the natives called *itica cullu, vettu cullu* or *kallu* and meaning brick stone or cut stone, he named, from the Latin, *laterite*, meaning brick.

Since then, by association, principally through location or use, numerous similar and sometimes dissimilar materials have been called laterites. The name, once assigned to a material, has been used as a substitute for properties and, as the properties of the various materials called laterite have often not been such as to lend them to the intended use, it is now a widely held misconception that the term is inaccurate and unsuited to definitive use. However, by referring again to Buchanan's derivation, we extract the Latin *later* meaning brick. In Western usage, brick refers to an earth material once pliable which by drying has been made irreversibly hard. Laterite, as the name implies, is a naturally occurring material possessing certain chemical characteristics which, after desiccation, hardens irreversibly. To assign the name laterite to materials which lack this property violates a literal interpretation of the definition.

As the geographic location of Buchanan's discovery was known, as well as the property of irreversible hardening, it was possible for those who followed him to determine the initial chemical make-up of the parent soil and postulate the environmental conditions under which the soil formed, the chemical compounds concentrated, and hardening took place. Bit by bit, a knowledge of soil formation and the hardening process was discovered.

A first requirement for laterite formation is a soil horizon of appropriate thickness, containing relatively high concentrations of iron and aluminum. Materials ideally filling this requirement are produced by the weathering of igneous and metamorphic rocks in a hot, moist climate; sometimes other soils are transformed. The original soil mantle typically contains feldspathic minerals, other silicates, and minor amounts of stable minerals. Intense chemical weathering is essential for proper preparation in the soil mantle. Subsequent to the transformation of the feldspathic minerals into clay, leaching and deposition occur in which iron and aluminum oxides remain after the removal of bases and combined silica.

Rock type plays an important part in the forming of laterite. In neutral or basic ferromagnesian-rich rocks, silicates are hydrolyzed and Fe_2O_3 and/or Al_2O_3 remain. Sialic rocks weather first to kaolinitic rather than lateritic soils.

Important in the forming process are the drainage characteristics of the soil. Well-drained soils formed from mafic rocks readily lose silicates along with bases. The soluble iron moves freely to the water table and accumulates, while the aluminum remains higher in the horizon. Poorly drained soils are nominally kaolin-rich, with lower horizons of black, iron-rich compounds if the barrier impeding drainage is a mafic rock. If it is sialic, the resulting iron is evenly dispersed and not concentrated, as it is on an impermeable barrier which deflects the direction of water movement.

Next in the forming process is a dramatic change in the environment, physical, such as the evaporation of much of the remaining water, or chemical, such as the reduction of groundwater temperature, ion exchange, or pH change. This results in deposition of the iron compounds and induration.

The placement, concentration, and type of chemical compound determine the physical properties of the primary concretion. These concretions usually form as nodules with a hard outer shell of ferrous material surrounding an inner core of softer or uncemented materials, frequently kaolinite. A crust thus develops which, in French-speaking Africa, is known as a *cuirass de fer* (iron breastplate), or *ferricrete*. Carvalho[2] refers to this ferrolitic cuirass as the first phase of laterite genesis.

After forming, the cuirass is joined to other adjacent cuirasses by a ferruginous detritic blanket. Beneath this, other accumulations of iron and aluminum compounds form which in turn are transformed to cuirasses and themselves joined into the relatively honeycombed structure of a hard skeleton containing softer inclusions.

These structures rarely disintegrate but remain in the characteristic laterite pattern throughout the undisturbed life of a soil horizon. It is now that the chemical balance of the salts is essential to irreversible hardening. The relationship of the three essentials—iron, aluminum, and silica—is shown in Table 1 using the method of identification reported by Jacques de Medina.[3] When the chemical requirements for irreversible hardening are met and the required change, physical and/or chemical, takes place, a laterite is produced.

Laterite may thus irreversibly harden *in situ* or may later harden when

Table 1. Identification of Laterite, Lateritic Soil, and Nonlateritic Materials (from de Medina[3])

$$K_r = \frac{\%SiO_2/60}{\%Al_2O_3/102 + \%Fe_2O_3/160}$$

	K_r
Nonlateritic soil	>2.00
Lateritic soil	1.33–2.00
Laterite	<1.33

exposed to air. Lateritic soils lacking the proper balance of constituents may become temporarily hard by desiccation but will subsequently soften when subject to moisture.

Continuation produces laterite of increasing depth as long as the required constituents are available for downward concentration and an anhydrous or chemical change takes place.

Usually laterite or lateritic soil, in place and protected, resists chemical weathering. Discounting extensive surface cover or dramatic erosion, laterites remain unchanged.

The occurrence of laterite in a system of sedimentary soil seems to require but slight modification of the conditions described for a residual soil resulting from igneous or metamorphic rocks. Clays and silts coming from weathered igneous or metamorphic rocks are most suitable. These impermeable feldspathic materials, if deposited upon sands and gravels, present a horizon which first permits a further weathering of the clays to kaolinite. Erosional sculpturing of the surface is required to leave kaolin deposits occupying ridges. From this positioning, groundwater percolating downward leads to the accumulation of iron and aluminum compounds. Often the initial cuirasses are formed within the permeable sands and gravels, provided the feeding stratum is thin. In a thick kaolinite development, it seems that cuirasses are formed low in the profile but not at the base. This would give rise to the possibility that the concentrated salts travel downward only so far before reaching supersaturation and deposition. The depth of initial cuirass formation is a function of the effective depth of desiccation or corresponding chemical change.

CHEMICAL CHARACTERISTICS

The results of an analysis of the chemical properties of laterites and lateritic soils from widely separated locations are given in Table 2. Each sample has been evaluated according to the test of de Medina[3] by equating the sum of the three measured constituents, iron, aluminum, and silica, to 100%.

Table 2. Chemical Analysis of Laterites from Various Locations

Sample location	Percent by weight: iron oxide, Fe_2O_3	silica, SiO_2	aluminum oxide, Al_2O_3
Andersonville Prk, Sumter Co., Ga.	27	16	0.1
Appling Co., Ga.	11	27	4.1
Kumba Division, Cameroon	9	31	8.0
Mamfe Division, Cameroon	32	11	0.4
Ellaville, Schley Co., Ga.	10	2.9	7.5
Thailand	29.5	47.3	10.2
Bombay, India	29.5	43.3	12.0
Cape of Good Hope, South Africa	34.5	25.7	19.0
El Salvador	11.6	34.9	26.1
Johannesburg, South Africa	37.9	31.8	15.0
Kumba, West Cameroon	29.2	25.2	24.8
Libreville, Gabon	46.3	10.0	27.1
Nairobi, Kenya	25.3	11.4	14.6

A summary of the observed conditions necessary for laterite formation first published by the author[4] in 1967 is listed below:

1. A soil mantle of appropriate thickness containing high concentrations of water-soluble iron and aluminum salts.
2. An underlying barrier which inhibits the downward departure of groundwater.
3. The introduction into the soil mantle of suitably accumulated rainfall or other water at proper temperature and acidity to dissolve aluminum and iron salts. (Monsoon seasonal rainfall is commonly associated with this parameter.)
4. A moderately low gradient to allow concentrated salt solutions to accumulate on a relatively impermeable barrier.
5. Changes, such as evaporation of water or the reduction of groundwater temperature, ion exchange, or pH changes, etc., which cause the salts to be deposited.

From this simplified sequence, the conditions of occurrence may be postulated.

Sialic or readily neutralized mafic rock was necessary for chemical change. The weathered mantle must have had a relatively impermeable base which diverted percolating water laterally rather than allowing it to continue downward. This requirement implies good drainage. In transported and sedimented material derived from erosion of weathered rock, the placed material must have been permeable and suitable for drainage or must have been of sufficient thickness so that the portion laterized would have been free of excess water. Initial cuirass formation required a high-density rainfall of 20 to 100 in. per yr occurring in a short period of time. Monsoon type rain is suggested.

ENVIRONMENTAL CONDITIONS

The introduction of the drying cycle may effectively occur subsequent to the formation of the lower horizon of cuirasses or may be delayed until several layers of cuirasses and the overlying and joining ferruginous detritic blankets have been placed. In laterite formed by cyclical groundwater concentration and subsequent hardening induced by desiccation, the rock structure has less continuity than systems laterized to the full extent of the available chemicals before final desiccation and hardening began. Furthermore, the depth of laterization is a function of the location of the first impermeable horizon. In a weathered horizon, in which the accumulation of the required chemical constituents has already taken place, induration at the surface effectively inhibits cementing compounds from being further transported downward. This phenomenon is exhibited in the surface hardpan laterite which renders ineffective further planting after the first few seasons on rain-forest land cleared for agricultural purposes. Rain-forest soils, long laterized, but protected from induration by the surface forest cover, harden in but a few seasons when the forest cover is removed, and marked soil desiccation takes place during the dry seasons. The surface hardpan laterite that troubles the new city of Brasilia is a notable example of this.

Laterite occurring in considerable thickness is the result of laterization of the soil to its full extent, followed by effective desiccation in proper sequence and extent to harden the materials somewhat uniformly. Though in extensive deposits the hardening probably occurs first at shallow depth and progresses downward, it is essential that sufficient permeability remains for groundwater to be drawn upward if induration is to progress below previously desiccated horizons.

Induration subsequent to cutting and removal of material is an artificial progression of the laterite-forming cycle. In this instance, blocks or chunks of "ripe" lateritic soil are removed and subjected to air drying, which produces induration. The materials to which Buchanan[1] referred were no doubt taken from such a ripe lateritic bank. In this case, the proper mix of chemical constituents has evolved within the soil horizon and the material is merely in a protected "deep freeze" from which it may be removed and indurated by either the sun or artificial means.

Voloboyev[5] has delineated the parameters of rainfall or groundwater induction by gravity and mean temperature which will produce laterite (Fig. 1). If the cycle of formation previously outlined is correct, the Voloboyev ranges must be subject to interpretation. These parameters are understandably inclusive, but occurrence in cyclical sequence is not necessary. Certainly rainfall or groundwater saturation might be followed by an extensive dormant period before induration is accomplished by the necessary drying effect of air temperature and drought. The classical cyclical application was probably meant by his suggestion, as it is this that is most frequently observed at the location of discovery of indurated deposits. However, in the deposits first reported by Buchanan, induration had to await temperature effects following exposure.

The rainfall suggested is no doubt meant to have taken place during a short period of time. The actual rate of appropriate rainfall is a function of the permeability of the soil horizon. The role of rainfall is to dissolve required salts and transport them in solution downward to the proper location in the mantle. Insufficient rainfall would fail to complete weathering of the feldspathic constituents. Examples may be noted in the Piedmont region of the

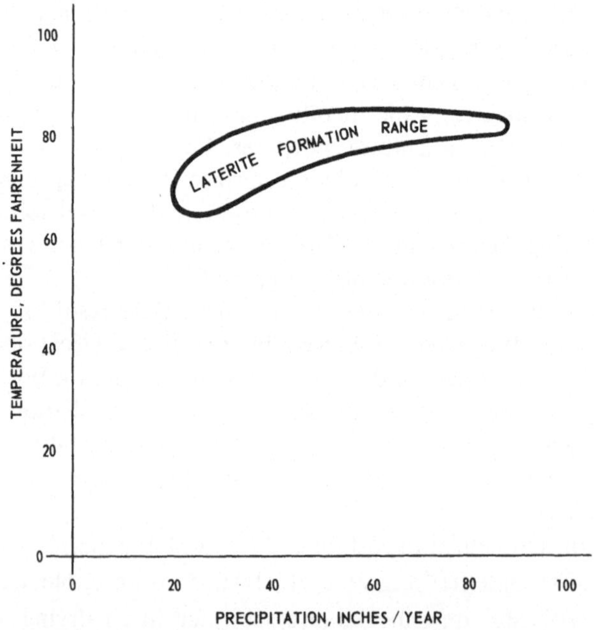

Fig. 1. Soil hydro-thermorange (after Voloboyev).

southeastern United States where proper constituents appear in a soil mantle, but have not been concentrated. This lack of geochemical concentration may be attributed to insufficient concentration of rainfall.

An excessive amount of rainfall would have produced weak, ineffective solutions of the necessary salts. Such solutions would have lacked ability to deposit and salts would have been carried away by the liquid. This may account in part for the absence of lateritic soil in the high-rainfall areas of the temperate forests of the southeastern Appalachians and the Olympic Peninsula in Washington.

According to the Voloboyev concept, 20 to 100 yr would provide sufficient time to effectively saturate a soil horizon. During this time the pressure head of groundwater must have been sufficient to cause transportation of iron and aluminum and then diminished so as to allow deposition. Excess rainfall which could not enter the horizon would have been carried off by surface drainage. Excess water must not have caused critical erosion of the upper soil mantle as this would have destroyed the feeding medium which was providing the iron and aluminum.

The seasonal interruption of rainfall is important for it allows groundwater saturating the horizon to *slowly* drain away and presents the conditions necessary for precipitation of iron and aluminum. Precipitation of these salts has occurred when a supersaturated solution could no longer hold its dissolved ions. However, if we are to presume that a characteristic downward temperature change produced the precipitation, we are confounded by the fact that stationary groundwater could hardly be expected to fall in temperature when the rainfall stopped. A simple explanation is that rainwater increased in temperature during its travel through the upper portion of the horizon since the surface of the ground was at a higher temperature than the air. During this travel it took the soil compounds into solution. As it moved lower in the horizon, the temperature fell sufficiently to produce deposition. However, the range of surface-water and groundwater temperature from the time of water induction until its departure at the base of the horizon is not sufficient to explain the observed accumulation of chemical compounds. Other phenomena were necessary in order to accomplish this.

Ion exchange or slight pH changes would have been sufficient, when coupled with a slight temperature reduction, to have produced the observed precipitation. The characteristic segregated configuration of the cuirasses suggests ion exchange occurring around a polar or orbital arrangement. This would explain the initial depositional positioning but would leave unexplained the precipitation of a detrital blanket or the structural connection of individual cuirasses. If orbital arrangement was the mode of occurrence of ion exchange, this orbit could have later served as a pole for further ion exchange. More simply, the proximity of high concentrations of precipitation followed a familiar drainage path through the laterizing medium which produced the lattice structure.

Pendleton and Shanasuvana[9] recognize two distinct forms of laterite: *vesiculan*, with a lattice structure of numerous dissimilar concretions joined together during formation, and *pisoletic*, exhibiting buckshot or pebbly concretions sometimes joined together with weak ferruginous cement.

CHEMICAL ACTION

Florentin and L'Heriteau[7] describe laterization as resembling kaolinization under climatic conditions characteristic of tropical environments. In their description, increased surface-water temperature raises the pH of the water and sesquioxides produced by hydrolysis precipitate as soon as the water reaches a strong alkaline level. Colloids of silica, however, become more stable in an alkaline solution and are transported through and away from the horizon of sesquioxide deposition.

The removal of the base from the upper horizon creates an acid environment which further accelerates the process and induces chemical changes replenishing depleted sesquioxides.

This continues until the weathered source rock or soil is completely modified. To explain this chemically, de Medina[3] provides the equations of kaolinization:

$$\underset{\text{Orthoclase}}{Al_2O_3 \cdot 6\,SiO_2 \cdot K_2O} + 2\,H_2O \longrightarrow Al_2O_3 \cdot 6\,SiO_2 \cdot H_2O + \underset{\text{Soluble}}{2\,KOH}$$

$$Al_2O_3 \cdot 6\,SiO_2 \cdot H_2O + H_2O \longrightarrow \underset{\text{Kaolinite}}{Al_2O_3 \cdot 2\,SiO_2 \cdot 2\,H_2O} + \underset{\text{Quartz}}{4\,SiO_2}$$

Contrast these equations with those for lateritization of kaolinite:

$$\underset{\text{Kaolinite}}{Al_2O_3 \cdot 2\,SiO_2 \cdot 2\,H_2O_3} + H_2O \longrightarrow \underset{\text{Gibbsite or hydrargillite}}{Al_2O_3 \cdot 3\,H_2O} + \underset{\text{Quartz}}{2\,SiO_2}$$

From these, we observe a continued rifting of silica from the sesquioxide Al_2O_3, leaving alumina available in relative freedom for concentration at the site of deposition.

PHYSIOGRAPHY OF OCCURRENCE

The work of D'Hoore[8] has resulted in the most simplified explanation of sesquioxide accumulation in soil. Figure 2 illustrates three frequent loca-

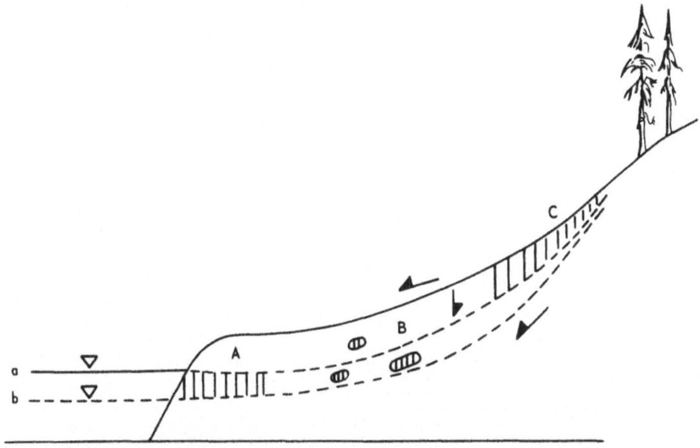

Fig. 2. Typical laterite locations.

tions of laterite and provides explanations of the phenomena which brought about accumulation.

In a river bank, **A** in which the water level varies from **a** to **b**, iron and/or other salts originating from river water seeps into the embankment, and is deposited, during flooding. Lowering of the water table produces desiccation and induration. At midslope, **B** salts migrating downward to the groundwater table, or along it, encounter conditions which promote precipitation. Individual cuirasses normally form. Further upward, **C**, lenticular crusts form near the top of a slope. Sesquioxides are precipitated from percolated water and form an impermeable base below the groundwater. Thereafter, all accumulation takes place along the base. Such a system results in a succession of detritic blankets which frequently outcrop at the surface, either through erosion of surface soil, exposing the laterite, or from continued feeding of sesquioxide solutions detained long enough to allow deposition.

Chapter 2

GEOLOGY OF LATERITES

PREPARATION OF SOIL MANTLE

Naturally fed laterites evolve in soils rich in iron and aluminum, the soils having derived from the *in situ* weathering of igneous or metamorphic rocks or from transported soils which have the same properties. A special case is laterite formed upon inert soils by enrichment of iron and aluminum oxides from an outside source.

The parent material to which Florentin and L'Heriteau[7] refer is orthoclase feldspar. Corresponding iron-rich minerals are usually present in suitable quantity and distribution in Precambrian hornblende gneisses and schists. Cambrian and later granites also possess appropriate constituents though they are normally replete with more coarse-grained quartz than is usually associated with laterite formed from a residual soil mantle over Precambrian rocks. Such rocks, later metamorphosed, sometimes dispose of their resistant quartz by accumulation in dikes and inclusions. Redistribution and redisposition of the feldspathic minerals is characteristic of the segregation of metamorphism. Zones of weakness which develop are often presented to the weather as upturned, receptive channels. Contrasted with these, intrusive rocks usually offer surfaces through which admission of surface water is gained only along fractures and the interfaces of inclusions.

The relatively violent reversal of internal stresses resulting from fluctuation of environmental conditions during weathering is not unlike that which takes place during metamorphism except that orogenic pressure is absent. Changes in temperature produce expansion and contraction in the surface rocks which separates masses along incipient fractures and interfaces between constituents. Water entering the rock, opens pores, destroys cementing materials, and softens less resistant minerals. When frozen, it produces interior forces which further commutes the rock. Desiccation reverses the flow in percolation channels and enlarges them, and, in addition, it aggravates contraction and tension, producing cracking and weakness.

LATERITIZATION

Pendleton and Shanasuvana[9] attribute most soil lateritization to the Tertiary period and principally to the Pliocene and Miocene epochs. Certainly, the geologic periods of high rainfall and accompanying drought, and absence of marked earth movement, are ideal for the formation of laterite. Lateritized soils which were formed in the Pleistocene have been observed by Cofer[10] and plotted by the author.[4] Lateritization of a soil horizon of positioned deposits of required salts is most easily postulated as a prehistoric geological occurrence requiring a great length of time. Yet, there is no evidence to support the hypothesis that lateritization must require millions of years to occur. Since interest in laterite is of quite recent origin, there have been few reports of its formation. However, the author has talked to many

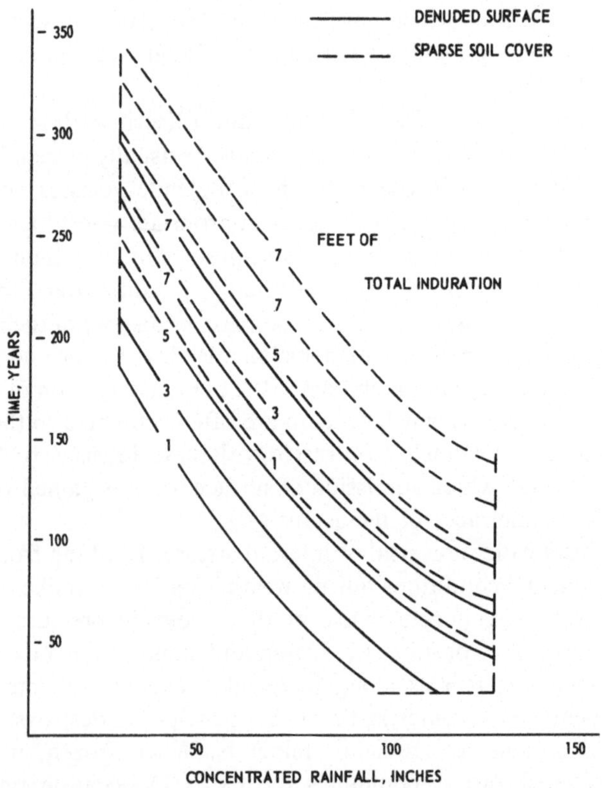

Fig. 3. Laterite formation relative to time and rainfall. The rainfall occurs in a 5-6 month period followed by a very dry period at 65°F mean temperature.

long-time residents of middle Georgia who identified roadcut faces which had changed from red clay to "strekeddy clay" in a period of fifty years. Excavation into these faces reveals only surface lateritization but illustrates that it may take place during relatively short periods of time.

INDURATION

Induration is thought to be a phenomenon limited to Miocene, Pliocene, Pleistocene, and Holocene epochs.

Capped Miocene laterite is discovered in bore holes in the Georgia coastal plain. As the cap (rock or impermeable clay) is also attributed to the Miocene, we may conclude that environmental parameters conducive to induration occurred during that epoch. Environmental conditions which could have produced induration are suggested for the Pliocene and Pleistocene by Cofer.[10] Buchanan[1] refers to Holocene induration occurring in what we may conclude is but a few seasons.

The author has postulated parameters of rainfall and cycles of drying (years) necessary for induration of protected deposits of ripe lateritized soil. These parameters (shown in Fig. 3) were arrived at by observing previously nonindurated deposits, solidified since the German timbering in Cameroon late in the 19th century (Fig. 4).

Fig. 4. A partially indurated laterite deposit in West Cameroon.

For the engineer, induration is the most significant property of laterite. An understanding of this occurrence is essential to his proper appreciation of laterite as a useful material.

Ripe lateritic material will indurate irreversibly when the water of hydration is mechanically removed from the sesquioxides, changing them to compounds lacking an affinity for water which might be induced at a later time.

Induration is an unexplained process in view of the published works which the author has examined. Certainly, it may be presumed that a high concentration of silica inhibits induration. The reasonable maximum of silica which may be present when induration occurs seems to be in the range of no more than 15%. Silica contents in excess of this value are sometimes present, as noted in Table 2. While the matrix materials appear indurated, they nevertheless soften by repeated water baths. This confirms the definition.

Fe_2O_3 is a superior cementing agent under most conditions and its presence in concentrations exceeding 25% further points to a state conducive to induration.

In a summary of the chemical properties of tropical soils reported by Professor Ta Liang,[11] an unpublished paper by J. E. Lukens[12] is cited furnishing ranges of the chemical composition of laterites. An extract of this summary, showing principal chemical composition, is given in Table 3.

Of these samples, the identifying ratio of silica to sesquioxides according to the equation of de Medina,[3] varies from nearly 0.0 to 3.42. However, Pendleton and Shanasuvana[9] report the identifying ratio of 31 samples of laterites taken from the building blocks of ancient Thai structures as an average of 0.79. The maximum value of the silica–sesquioxide ratio for *induratable* material is believed to be usually no greater than 2.5. *Values in excess of this for materials known to be irreversibly indurated may be attributed to an addition of deposited silica subsequent to induration.* Examples of this are recorded in Table 2.

Ferric oxide may well be induced into the system in a form other than that of its final indurated state. The removal of absorbed carbon dioxide

Table 3. Chemical Composition of Laterites

Chemical composition	Average, %*	Extreme, % low	Extreme, % high
SiO_2	16.6	0.6	49.8
Al_2O_3	20.1	2.7	38.7
Fe_2O_3	47.5	20.7	81.9

* 54 samples

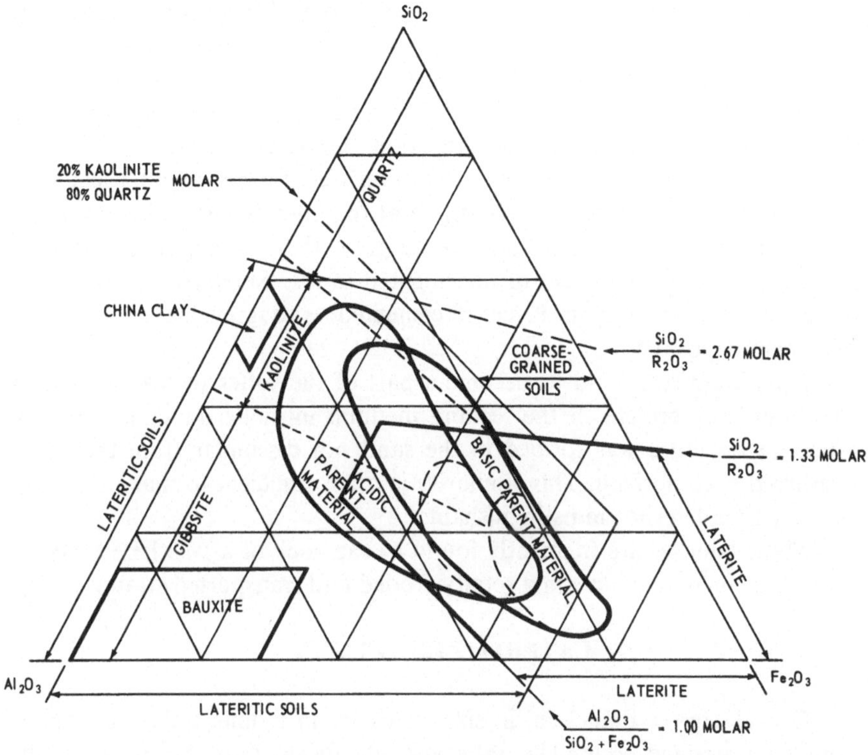

Fig. 5. Composition of laterite and lateritic soils (after Lukens).

during desiccation could produce the necessary valence change to catalyze hardening. However, ferric oxide is abundantly found in lateritic soil and in nonlateritic soil in the same state as in laterite. The factor most reasonably responsible for hardening is thought to be the degree of iron-salt concentration at the time of desiccation.

The relationship of the three basic composition chemicals of laterites and lateritic soils reported by Lukens(12) is shown in Fig. 5. The percentage of each of the three essential constituents must be related to a total of 100 when applying this means of identification. The exclusion of trace minerals, water, and relatively inert materials such as silica sand is required.

SEDIMENTARY LATERITES

Laterite forming in relatively fine-grained transported sedimentary materials was mentioned earlier in this chapter. These soils have been transported but a short distance and have retained all of the properties essential for laterite formation, similar to property retention in residual mantles of weathered rock. Laterites considered in this section include only those produced in sand strata in which the chemical feeding elements are above. Such conditions would have resulted from a thin sand horizon underlain by an impervious base and overlain by a stratum containing the iron and aluminum salts necessary for lateritization. Upon the introduction of the salts in solution, in this situation, concretions were formed about select sand grains and a thin indurated layer began at the interface with the lower impermeable stratum. As the concretions grew they were joined together, frequently leaving interstices of uncemented sands. Similarly, the thickness of the usually continuous layer of cemented sand grew as concentrated salts were deposited.

Aluminum salts tend to become a part of the softer inclusions, and if aluminum was present in the feeding medium in sufficient concentration, a laterite structure was formed in the sand not dissimilar from that of a weathered rock horizon. This similarity is often sufficient to make the sedimentary-type laterites indistinguishable.

These laterites are frequently found as cap rock of a weathered coastal plain or as mere remnants in a totally eroded and transported stratum.

LATERITIZED GRAVEL

Gravel greater than 1 in. in size lateritizes in a quite different manner from finer grained soils. The salts are introduced into the gravel itself. Starting from the middle of a piece of gravel, they give the material a reddish hue, continuing outward and reaching the surface after prolonged deposition. In close proximity, the surfaces of two pieces of gravel may trap sufficient iron salt to cement the pieces together. If this occurs in a closely packed or well-graded stratum, connection proceeds as with the joining of cuirasses in a fine-grained system. Normally, however, aluminum is rarely found in high concentration in deposits of large-sized cemented gravel.

Individual pieces of gravel are less hard following lateritization than in their initial state. The cementing blanket, however, is quite hard, though usually more brittle than laterite from a sandy-soil system.

Chapter 3

FINDING LATERITES

WORLD DISTRIBUTION AND MAP

At some time laterite has existed everywhere that the parameters for its production have prevailed. The frequently changing temperature and rainfall conditions throughout geologic history have presented the environment necessary for accumulation and induration of iron salts to much of the earth's surface within the boundaries of the temperate zone between 35° N and 35° S. As iron, aluminum, and silica are sufficiently abundant in surface rocks, it was only necessary that the proper concentration be present when lateritizing conditions existed. The environment which produces laterites is severe, and severity is not usually of long duration in geologic terms. Incomplete weathering when a laterite-inducing environment prevailed seems to be the principal factor which would have prevented laterite formation from rocks containing the necessary elements. Complete weathering is necessary to free the elements and create the sesquioxides which must be present. Insufficiently weathered igneous and metamorphic rocks of the Appalachians are an example of incomplete weathering during the periods of necessary conditions.

In addition to those places where it has been found, laterite was once present at those locations from which it has since been removed by erosion or other forces. It may also be expected at those locations where, though it has not been discovered, there have been environmental conditions suitable for accumulation and induration and the necessary soil constituents for lateritization.

In preparing a world map of laterite location, it is necessary to include all areas within the environmental limits which contain residual and transported soils with sufficient iron for induration.

The world distribution map in Fig. 6 shows areas of known laterite occurrence, historic (presently lateritizing) and prehistoric, and areas in which it is expected that laterite will be discovered. Because of the requirement of continental reach, certain generalizations have been made which will not appear inconsistent to those familiar with regional geology. As this map

Fig. 6. World distribution map.

tends to be inclusive rather than exclusive, a knowledge of regional soils is essential for local interpretation.

For those persons who would use this map as a guide to the location of engineering materials, attention is drawn to the specifics described in the sections which follow.

IDENTIFYING AND CLASSIFYING KNOWN SOURCES

The engineer who would build in laterite areas needs first to have a thorough understanding of local practices, and location and usefulness of local materials. Almost universally, where available, iron rock or iron soil is used in one form or another in engineered construction. In a new area, it is well for the newly arrived engineer to be sparing in pronouncements and eager to observe.

Most likely a keen eye will reveal subgrade fills, base course, and wearing surfaces made of laterite or lateritic soils. Retaining walls, dam riprap, and building stone may be of iron rock. Stockpiles of building stone or surfacing materials will frequently be seen about public works yards or at strategic locations adjacent to construction sites. Loaded dump trucks will reveal fill and surfacing soils.

Discrete inquiry will often result in the disclosure of the source of these materials. In noting the source, it is important to relate local practices to the actual place of use of those materials derived from a certain source. A few days of well-timed inquiry and observation will be invaluable during later materials searches.

A visit to construction locations is essential before actually visiting and assaying quarries or borrow sites. If possible, the engineer should firmly state his desire to visit construction sites and maintenance areas to view local practices firsthand, understand them, and relate local knowledge to his purposes.

In viewing construction sites, the engineer should begin his catalog of materials characteristics, location, and use in a systematic manner lest he reach improper conclusions. Preferably, his assembly of data should be a narrative record, as tabulation often fails to disclose information essential to evaluation.

Essential to the record is a separate study of each site with location, physical characteristics, and areal dimensions. The date of the deposit discovery and the results of material tests prior to first use are important, as is a record of routine periodic tests performed during operation.

An inquisitive study is necessary to identify the site of use of the material. As each of the sites of use is visited, a separate report should be prepared for each showing location, length of haul, method of use, sequences of placement during construction, and attention to detail and workmanship of these operations. (For example, it is beneficial to know the gradation of the laterite rock as it arrived at a surfacing site as well as the method and degree of compaction of the crushed materials.) Care and attention to maintenance since construction are further keys to evaluating material behavior.

In studying a borrow or quarry, the material in the pit should be evaluated for uniformity, both vertically and laterally. Materials from all construction areas should be collected and compared with the materials currently being produced in the quarry.

By judicious selection, samples of the quarry rock should be acquired and subjected to tests (field and laboratory, if necessary) in order to identify the materials for degree of induration, compaction, characteristics, and site moisture. The extent of the remaining deposit should be estimated. Field and laboratory tests which may be used to identify laterite and lateritic soil are suggested in the last section of this chapter.

Upon completion of the study of known sources, the engineer should evaluate the results. This evaluation is of utmost importance in the future selection of construction materials.

In evaluating a source, the intended use and its effectiveness should be of first importance. If the effectiveness of a material is judged less than desirable, the reason for this failing must be found.

This study deals with laterite, an irreversibly indurated material. Care must be taken that the name laterite is not being assigned to lateritic soils which harden slightly on drying but soften when subjected to water. Another frequent misunderstanding arises concerning a laterite quarry which displays a cap or face of obviously well-indurated rock, and from which materials have been taken and placed which have misbehaved under use. Most often, it will be discovered that the materials actually used from the pit were not indurated, but easily excavated soils. Often such sites of the easily quarried but poor materials were worked and were abandoned when the hard, useful materials were encountered. It is often difficult to reconstruct this sequence of events. Testing and comparison of used material and remaining quarry rock is sometimes necessary to discover the discrepancy.

From an objective evaluation of purposes, effectiveness, and characteristics of known borrow locations, the engineer should conclude his appraisal of local construction practices. These conclusions should form the basis of the first selection of desirable materials and rejection of the undesirables. With this evaluation, the engineering geologist should relate the mode of occurrence and location characteristics of all sources. These often elemental, but very important, terrain keys will prove timely in later materials searches.

Quarry sites, if properly evaluated, can prove to be the most useful guide to the location of other hidden materials.

The newly arrived engineer is warned against casting criticism toward local construction practices or material uses. His effectiveness in accomplishing his purposes will depend not only on his knowledge, experience, and ability to translate this knowledge to immediate conditions, but also on his ability as a diplomat. No one likes to be told that his historic methods are totally wrong.

The engineer should remember that if his talents had not been needed, he would not have been summoned in the first place, and that he was called not as a critic, but as a performer. Performance will entail recognizing, accepting, and using good local practices and avoiding poor ones. The latter often will require reeducating local officials and this can best be done by proceeding slowly. For the time being, at least, the engineer should keep his evaluation of local conditions to himself, for his own future use.

DEFINING AREA OF NEED

In most locations where laterite is to be used as a construction material, its use will either be intensive (as in dams, buildings, or town or city streets) or extensive (as a base course or wearing surface for roads or railroads).

For an intensive requirement, sources should be explored which take advantage of the most expeditious and effective means of getting the material to the location of use. Transportation to the location of use should be extended only from the maximum desirable distance for importation. Relatively inaccessible areas should be examined with an awareness that access roads must be built.

For extensive use, a band of terrain immediately adjacent to the construction site is of first and foremost importance. Laterite construction is sometimes restricted to remote, heavily forested areas plagued with intensive rainfall and poor soil. It is difficult enough to build a road or railroad in these areas without the added burden of building and maintaining lengthy haul roads requiring inordinate quantities of precious base and surfacing materials and unproductive use of personnel and equipment for maintenance during wet periods.

It is better to use the road under construction as the haulage road and expend material, equipment, and time upon it. Following this logic, it is desirable that laterite pits be immediately adjacent to the newly laid route or beside an existing and easily maintained nearby road. For borrow pits requiring construction of haulage roads, a half-mile distance from an existing road is excessive. Present-day evaluation of suburban haulage and overhaul have little place in the thinking of an engineer building roads in the rain forest.

An engineer must understand thoroughly the capabilities and limitations of construction techniques before he selects areas in which he wishes the geologist to find materials of construction. Furthermore, he must communicate this intelligence to the geologist so that he too may realize the importance of locating "close by" materials.

A useful exercise in defining acceptable borrow locations is to set the existing road net on a relief map of the area and to weigh this in terms of trafficability during construction. Thorough engineering reconnaissance of the road net is necessary, together with an accumulation of information from local residents concerning road conditions during various seasons and under various traffic conditions. The tentative or actual route should be marked on the map and if it is to be considered useful for transport of borrow material, construction sequence should be carefully considered. Of particular importance when considering the use of a road under construction, are the apparent obstacles, such as bridges or deep fills, which may hinder or prevent the linkage of easily constructed sections.

TERRAIN KEYS

Cofer,[10] Carvalho,[2] Liang,[11] and de Medina[3] have described laterite occurrence which fit a general pattern. The locations of laterite deposits in West Africa which the author has seen and discovered conform to the general terrain pattern predicted by these authors.

In the normal, flourishing rain forest, ripe laterite may be expected:
- In the upper third of well-drained ridges.
- Beneath sparse vegetation on hilltops, slopes, or tabletops.
- Beneath locations of grade change in forested, ungullied drainage slopes.
- At rubber plantations.

Indurated laterite may be expected:
- Beneath long-denuded surfaces of well-drained ridges.
- At the extremities of isolated tabletop prominences.
- At abandoned farms.

Cap-rock or ironseam laterite may be expected:
- At locations of minimal vegetation.
- Beneath the surfaces of poorly drained swails or depressions.
- Within high-water range of iron-bearing streams.

In prehistoric rain-forest areas now grasslands or deserts, laterite both indurated and ripe may be expected:
- On narrow ridges near the upper-third surface.
- Beneath the entire surface of isolated tabletops.
- As remnants along eroded surfaces.

Liang[11] further comments on the high occurrence of laterite at golf courses, burial grounds, and areas having termite hills. The affinity which termites, golfers, and the buriers have for indurated materials is immediately apparent. Moderately high-and-dry ground is necessary for all and they have selected the areas which the environment also selected to lateritize. A bit of imaginative thought directed to similar land uses can result in many other locations in which to search. Abandoned airfields, particularly in deep forest, may have been cleared long enough for induration. The local fort and its adjacent parade are likely to be laterite. The hill and ridge portions of forest road right-of-ways are nearly always laterite and in road relocation, the old road bed could prove an excellent source of borrow materials. The face of steep slopes or escarpments are usually poorly vegetated and likely to be lateritized.

In most forest locations, the seeker can establish a relationship between vegetation and laterite. Though crude, it will be of untold benefit in deciding whether or not to hack a road through the bush and drag a drill rig several hundred yards off the road.

At known deposits, the types and density of older trees should be noted within the immediate terminal drainage area (intermittent water course) and along slopes between water courses and the lower extremities of the borrow location. An unfavorable departure from the number and type of trees at another, similar terrain feature would render highly unlikely a laterite deposit on the up slope or ridge top.

Iron-rust water in pools or along banks of intermittent streams shows an excess of iron-bearing water in the headlands and may be traced to a nearby laterite or ripe deposit of laterite yet unhardened. Intense search is warranted when iron rust is discovered.

AERIAL PHOTOGRAPHIC STUDY

A most comprehensive study of airphoto interpretation for the location of laterite has been conducted by Professor Ta Liang.[11] From his study, and recently published reports of the Aeronautical Chart and Information Center of the U.S. Air Force and the U.S. Geological Survey, airphoto coverage for the laterite areas of the world has been summarized (Table 4).

An engineer or geologist interested in locating laterite in a specific area should address his request for aerial photographs to the agency most likely to have photographs on file. Experience has shown that locating and obtaining photographs of a remote area may be difficult and time consuming, and the engineer or geologist who suspects he may have need for laterite or who may be charged with its location in an impending assignment should endeavor to secure airphotos before arriving in the field. Usually a commission is known of long before departure, and credentials for authorization to secure

Table 4. Available Airphoto Coverage

Agency	Areas
U. S. Department of Agriculture	Most areas of the United States.
U. S. Geological Survey	Information source of airphoto coverage of entire world. USGS has a complete file on location and required means for securing photos throughout the world.
U. S. Army	The U. S. Army performed extensive aerial photography during World War II and has continued this mission. It has extensive declassified coverage which may be available for engineering use.
U. S. Air Force Aeronautical Chart and Information Center	The ACIC has extensive airphoto status maps for each country and will release these data to those who have the authorization of the local government.
British Directorate of Overseas Surveys	British overseas territories, Commonwealth countries, and former British colonies.
French National Geographic Institute	French overseas territories and former French colonies.
Spanish National Geographic Institute	Spanish overseas territories and former Spanish colonies.
Portuguese National Geographic Institute	Portuguese overseas territories and former Portuguese colonies.
Organization of American States	Airphoto information of all Latin American countries.

airphotos should be obtained at the first opportunity from the sponsoring government; the client should request the release of the airphotos, if a private enterprise. Once government approval has been obtained, the required photographs, if available, may be secured most promptly by visiting the agency which has the negatives on file. A certain urgency is demonstrated by a personal visit.

If the extent of an engineered project is known, it is usually wise to obtain route photographs of the entire area together with coverage of outlying areas adjacent to existing roads or railroads.

Appreciation of the desirability of laterite as a material of engineered construction has prompted many recently published works concerning the location of this material by means of airphoto interpretation. Liang,[11] Caiger,[13] Brink,[14] Holden,[15] Loureiro,[16] and Mountain[17] have contributed noteworthy procedures. The Cornell University Navy Report Land Form Series[19] illustrates the airphoto patterns of most common rock formations. From these reports, the summary for laterite identification which appears in Table 5 is derived.

Table 5. Summary of Airphoto Identification

Materials	Topography	Drainage	Erosion	Grey tone	Vegetation	Remarks
Laterite (crust)	Cap rock, flat hilltops, ledges on slopes	Slight	Little (rockfalls on edge of cap)	Light	Grass or low shrubs	Characteristic thin, light-grey line within darker tone
Laterite (gravel)	Flat hilltops to rolling terrain	Slight	Little (gullying on slopes below gravel)	Light	Grass or low shrubs	Light-grey mass above slightly darker grey eroded area within darker tone
Laterite (hardened on exposure)	Rolling terrain with flat valleys at high to intermediate slopes	Slight	Little	Light	Plantations, swidden agriculture, stunted trees, light-grey foliage	Shows borrow pits, abandoned farms, airfields without vegetation, denuded golf courses, parades, etc.
Lateritic soil	Rolling terrain at high to intermediate slopes	Varies from moderate to slight depending on degree of laterization	Varies little in more advanced stage of lateritization	Light to medium	Plantations, swidden agriculture	Termite hills

As a first step, the laterite hunter would do well to observe various laterite deposits in the field and compare these with their aerial photographs. Subtle terrain indications, identifiable from photographs, will be noted. Some, which have been frequently reported by the referenced authors, are listed below and illustrate things to look for in systems which might either possess or clearly lack laterite deposits.

Airphoto Keys to Laterite Location

Evidence of landslides points to youth and instability of a soil system and precludes formation of laterites.

Flat hilltops with surrounding thin light-grey lines are positive identification of laterite cap rock exposed at edge.

Recent transported soils are too young to form laterite.

Residual soils derived from igneous and metamorphosed igneous rocks are most likely to be lateritized.

Rolling terrain suggests a mature system more likely to be lateritized.

Flat-bottomed valleys denote recent alluvium lacking in laterite, except where laterite might be expected at the edges of valleys bordering the feet of adjacent hills.

Rugged terrain suggests a young system in which lateritization has not taken place.

Featureless, forested terrain suggests an ancient system not lateritized.

Features showing light tones in an otherwise flat system suggest erosion-resistant laterite deposits (particularly associated with sedimentary-soil laterites).

Lack of surface vegetation in a forested area strongly points to surface laterite.

Evidence of swidden (shifting) agriculture in a forested area strongly points to soils which rapidly lateritize on clearing.

Special note must be given to laterites within the tropical rain forest. In these areas, rainfall is so plentiful that trees will grow on most soils, even thin lateritic soils overlying laterite. Trees derive nutrient from their own droppings and the root system often induces sufficient moisture into the soil to inhibit induration. However, when the surface is denuded, an area becomes lateritized beneath the surface and the vegetation which subsequently covers the surface will live a tenuous existence never attaining the stature of the surrounding forest. The correlation of tree types and heights with laterite location is a subject yet to be properly studied. The author has noted that the foliage of trees at laterite deposits is of a lighter shade than similar trees on soils yet to be lateritized. This, however, might prove a misleading means of identification due to the variety of colors and

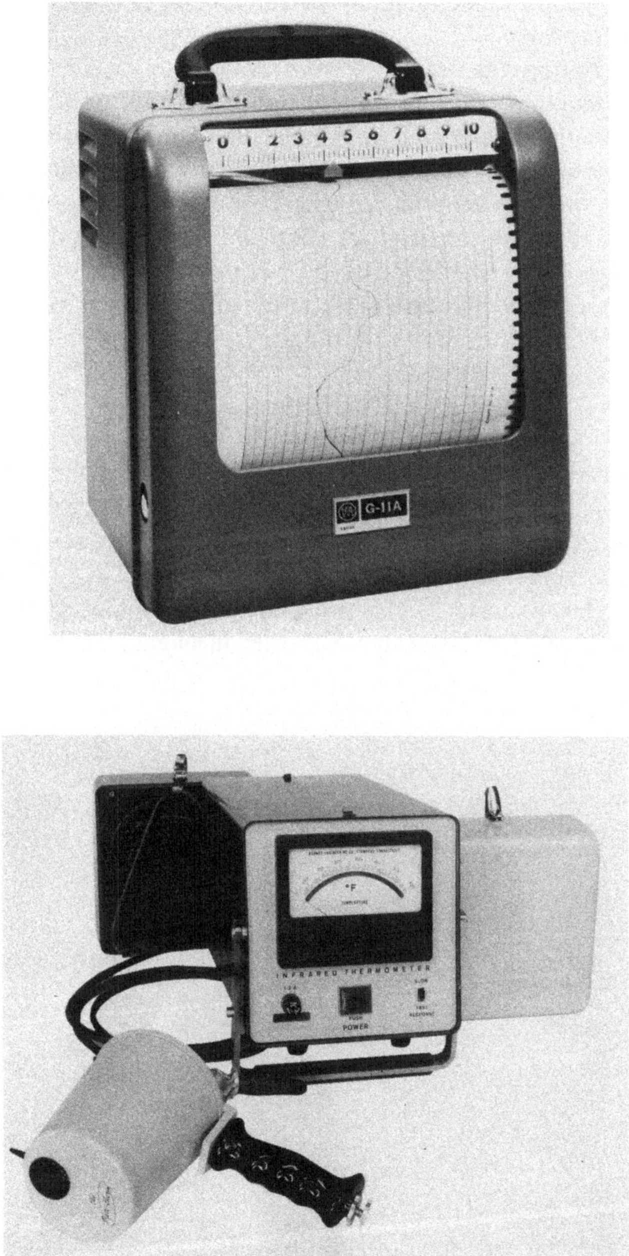

Fig. 7. Barnes infrared aerial scanner (bottom) with Varian strip recorder (top). (Photographs: aerial scanner, courtesy Barnes Instrument Company; strip recorder, courtesy Varian Aerograph.)

shades of foliage which are found in the tropical rain forest, but unassociated with laterite deposits.

At this writing, the use of a comparative study of panchromatic and infrared photographs taken on the same day suggests the most promising method of identifying hidden terrain anomalies. Drainage patterns and certain tree types stand out well in infrared photographs. Surface-temperature variations as well as variations in temperature between various trees may signify areas requiring intensive photographic study and possible ground reconnaissance. Small scale, good quality panchromatic photographs of areas of special interest may disclose a more meaningful perspective of land forms, soil types, erosional scars, and even surface soil thickness.

INFRARED AERIAL SCANNING

Coupled with the use of infrared photography, aerial scanning with an infrared sensing device offers a promising method of disclosing laterite in heavily forested areas. There are several infrared scanning devices currently available. The author is acquainted with the successful use of the Barnes Model IT-3, shown in Fig. 7. This assembly, priced at approximately $1000 and weighing about 30 lb, is easily transported in a light plane and is suitable

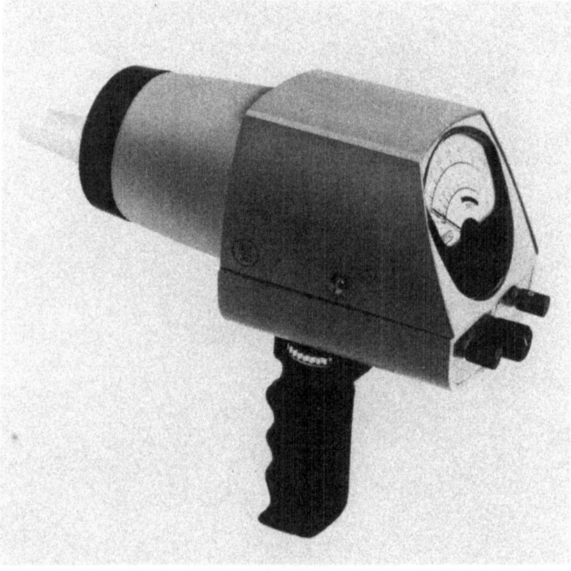

Fig. 8. Infrared field thermometer Model PRT-10. (Photograph courtesy Barnes Instrument Company.)

for low-level reconnaissance. Fitted with the most useful 3-deg field of vision, the scanner will identify thermal variations of the surface to which it is directed of 1° F. A strip recorder similar to the illustrated Varian Model G 11-A in Fig. 7 plots a curve of surface temperature for the field traversed in flight.

A recently developed instrument is the Barnes Model PRT-10, an infrared field thermometer more appropriate to ground reconnaissance than the larger Barnes unit (Fig. 8). This device may be used effectively for intensive study of suspected locations to more closely identify surface-temperature variations.

A suggested procedure for locating laterite deposits by the use of infrared sensing would require the identification of temperature variations at the surface of known laterite deposits and relating these to variations at identifiable sites located during aerial scanning. This method of identification has not as yet been used successfully in the location of laterite. A description of its utilization in the plotting of surface isotherms is given by Cheney and Richards.[19] The techniques described in their paper may well be applied to laterite location through imaginative application of the device and recorder.

EXPLORING SUSPECTED LOCATIONS

Explorations* directed toward discovering laterite rarely are of like finesse and sophistication as those marshalled for the discovery and evaluation of most other materials. Because of heavily forested and remote locations of laterite deposits and the tortuous road net to these locations, mobile, versatile drilling equipment is rarely suited to the task. Furthermore, because of the very nature of the intended use of the material sought, the project is usually budgeted low and imagination in exploration is more important than brute force.

The engineer or geologist seeking to explore for laterite for use as a construction material must select exploration equipment that realistically fits the particular task at hand. Often, there is exploration drilling equipment available in the local public works department and this should be utilized within its capabilities before other more suitable equipment is requisitioned. Presuming the seeker is lucky enough to get a relatively free choice in selecting exploration equipment, there are certain guides which are helpful in choosing the drills and other devices to be employed.

---

* "Exploration" as used in this volume means, in the geologic sense, both finding a deposit and estimating the quantity and quality of a known deposit. The engineering usage of these terms would be preliminary exploration and detailed investigation.

Guide to Choosing Exploration Equipment

The equipment should be suitably mobile to be moved to location in economical time.

The equipment should be capable of penetrating and retrieving suitable samples of the deposit of interest.

Samples retrieved should be in a state suitable for identification and physical as well as chemical testing.

The initial cost of the equipment, its cost of operation compared with its rates of progress, its cost of maintenance, and its final value must be commensurate with the cash outlay that may be reasonably ascribed to the task.

The rates of drilling progress and the time lag between breakdown and repairs should be appropriate to assure that the equipment will perform its intended task within the allotted time.

The availability of the equipment, its delivery time, and the delivery time of unanticipated repair items must fit the exploration schedule.

The operational procedures must be such that untrained, unsophisticated personnel can perform them successfully without damaging the equipment.

Generally, suitable laterite samples can be obtained from drill holes 4–6 in. in diameter. Smaller holes are rarely attainable through partially indurated deposits nor do they allow for recovering unfragmented cores suitable for testing. Larger holes require the expenditure of unwarranted effort for the return achieved.

Carbide-toothed augers are usually suitable for drilling even the hardest laterite. Hawthorn and similar roller bits are also suitable, though these are often associated with drilling methods that fail to deliver the spoil in identifiable, usable form. Untoothed augers are suitable for soft laterites or lateritic soil but they often refuse on hard laterite.

A continuous flight auger, operated in the dry, is probably the most suitable tool of laterite exploration. A short-length (one to three flights) auger extension above the bit is generally good for recovering spoil though it requires pulling the rods after each run. Rotary-wash drilling is time consuming for shallow depths of exploration, difficult to mobilize and perform in a remote location, and unsuited to necessary sample recovery. The tenuous and crude churn drilling is less suitable.

A sawtoothed core barrel can be effectively employed to retrieve drilled material but should be equipped with an automatic plunger to expel the retained spoil if the operation is to be expeditious. Drive samples are rarely suited for physical testing as the driving of a relatively small diameter (3–5 in.) sample usually destroys cuirasses and the laterite structure. The best core samples are obtained by the use of a carbide- or diamond-tipped, single- or double-tube core barrel. The Dennison sampler is too sophisticated for use in remote locations.

Mechanical equipment to drill well-indurated laterite should have a drawdown force of 2000 lb for a maximum 8–in. diameter drill bit. Lesser force may be suitable for the core barrel provided there is not an inordinate dissipation of drilling force due to side friction.

A suitable drilling machine which the author's firm has employed for laterite exploration in Cameroon is the Giddings Model GSRT, shown in Fig. 9. This machine is light (2200 lb), trailer mounted for fast movement by

Fig. 9. Giddings drilling machine. (Photograph courtesy Giddings Machine Company.)

road, and can be maneuvered by hand with relative ease off the road and through moderately open rain forest. Its drawdown of 1200 lb is at the minimum for hard laterite but its versatility, simplicity of construction, and ruggedness somewhat negate this deficiency. It operates most effectively with a continuous flight or short auger attached to the drll rods. It effectively drills with a core barrel, though slowly in hard laterite. The latter procedure requires some modification of the equipment but this can be accomplished easily in a field machine and welding shop. The total cost of the unit is approximately $4000.

A huskier machine is the Acker Model MP 100 Drilling Machine which, being truck mounted, is limited in mobility. However, if suitable earth clearing and smoothing equipment is available, the increased capacity and force of this machine more than outweighs its size, weight, and lack of mobility. The total cost of the complete truck-mounted unit is approximately $13,000.

Ponderous skid-mounted machines as well as truck-mounted, rotary-wash units have little place in laterite exploration.

Other machines which the author has noted to have many of the required capabilities are Acker AP Auger, Acker Model SP68, Mobile Model B–27, Mobile Model "Apache," Sprague & Henwood Model 30, Mobile Model Explorer, Penndrill Model "B" Testborer, and the Acker Model LD Core Drill. The reader is referred to the manufacturers for details on these machines, listed below.

Giddings Machine Company, 401 Pine Street, Fort Collins, Colorado.
Mobile Drilling, Inc., 960 North Pennsylvania Street, Indianapolis, Indiana.
Sprague & Henwood, Inc., Scranton, Pennsylvania.
Acker Drill Company, Inc., Box 830, Scranton, Pennsylvania.
Pennsylvania Drilling Company, 1205 Chartiens Avenue, Pittsburg, Pennsylvania.

A suspected location should be explored in a manner which will disclose only the necessary information. Laterite deposits are usually continuous throughout a lateritized area. Peripheral explorations will confirm the areal extent of the deposit as well as the edge thicknesses. Interior borings are desirable to confirm the usually greater thicknesses of the deposit at its middle. (This does not apply in valley side slopes where laterite is thickest at the edges.)

The location of borings around the periphery of a suspected deposit should coincide with the outer limits of possible borrow. In tenuous, forested, hilly locations, the extent of clearing required and the accessibility of terrain forms for borrow operations must be evaluated as well as the location of the deposit, haul distances, and the amount of material required.

The sequence of engineering design and construction usually follows from need and concept, to reconnaissance, to preliminary design, to final design plans and specifications, to bidding and award of the contract, to construction with professional surveillance and inspection. During reconnaissance, the materials of construction are usually selected and, if laterite is to be a material of construction, its presence in sufficient quantity is normally ascertained by geological mapping.

During preliminary design, the engineering properties of materials are normally measured at proposed borrow locations by a minimum number of exploration borings. When final design is begun, there is a general idea of the locations of prospective borrow. These locations are evaluated relative to possible use and general parameters of material needs can be set. These estimates will disclose the desired extent of a material at a given location. The exploration program should confirm the desired amount of materials.

Once the engineer has established the quantity of material required and the geologist has predicted the configuration of the deposit, together they can estimate the limits which can be reasonably worked. The outer exploration borings are drilled along these limits. Once the lateral extent of the deposit has been assured, interior borings of sufficient number to satisfy the geologist should be added.

Exploration borings should disclose the extent of waste overburden and underlying hard materials. At the outset of an exploration and testing program, and before field personnel become familiar with the field identification of suitable irreversibly indurated material, very frequent sampling is necessary. A sufficient number of loose laterite samples should be sent to the laboratory for identification. As moisture content is not significant in laboratory study, gallon cans with secured tops are ideal for containing such samples. This volume of material is sufficient for a CBR (California Bearing Ratio) test as well as numerous other laboratory studies. A 30–lb bag sample should be obtained from each boring for compaction tests. A few cores of material should be obtained for special laboratory testing such as unconfined-compression, triaxial, and consolidation testing.

FIELD AND LABORATORY IDENTIFICATION

Considerable attention has been given by soil scientists, geologists, and engineers to field and laboratory identification which would serve to differentiate permanently indurated materials from those which soften under load and/or moisture conditions. Frequently, the methods derived are to be used in a particular construction or within a limited geographic area and are contained solely in job specifications, receiving only limited distribution to those not connected with the job. These have served as the basis for regional or area specifications prepared by public or military works departments. A typical example of regional specifications prepared along these lines is that developed by A. Remillon, discussed by de Medina.[3] Criteria are cited for grain-size distribution, compaction and strength characteristics, the resistance-to-abrasion property, and the property of absence of softening during inundation. The field engineer or geologist who must find laterite can be somewhat confounded by all of these standards which necessitate extensive laboratory testing.

Recently developed as a part of the course literature of the Engineer School, United States Army, Ft. Belvoir, Virginia, is a chapter entitled *Laterites and Lateritic Soils of Vietnam*, designed and written to serve the user directly in the field, both as to field identification and subsequent use of laterite and lateritic soils. As this work may not be readily available for reference, it is included in its entirety as Appendix A, and is printed with the permission of the Engineer School.

An examination made by the Engineer School of the physical properties of two types of laterite and lateritic soil discloses moderately higher specific gravity for the laterite (average 3.13) than for the lateritic soil (average 2.92). A MOH's scale of hardness of about 3 is present in the harder inclusions of the laterite, while such values are not attained by any portion of the lateritic soil. Induration to this hardness is rarely present in lateritic soil except in the pebbles, which are true laterite in the form of primary cuirasses. There is a total absence of gravel-sized (primary cuirasses) material in lateritic soil, while this gradation is present in appreciable quantity in the laterite distribution.

Pebbly laterite should be used as a construction material with recognition of the probable behavior of the matrix of fines, unless it is intended that this fraction be removed or it is of insignificant quantity in the material.

The method of evaluating the degree of lateritization of the fines which is described in Sections 9a and b of Appendix A is the most direct method of field or laboratory identification known by the author. A ball of the minus #40 sieve material is submerged in water. If it is observed to disintegrate quickly, it is concluded that the matrix material is not significantly lateritized. If it disintegrates slowly during a 24-hour soaking, it is judged to have been sufficiently lateritized to possess the constituents for final, irreversible induration when employed as a construction material subject to drying, as in a road or airfield surfacing.

A more refined laboratory identification of the degree of lateritization involves repeated soaking and drying of the ball of material to test if hardness is progressive through a repeat of the process.

Field identification of laterite indurated to a continuous mass may follow the same procedure. Materials which resist excavation by a hand shovel and which ring when struck can be considered as indurated if they pass the soaking test described above. The author has observed Cameroonian materials which appeared to be indurated but which disintegrated on soaking. The degree of difficulty in excavating is not sufficient to differentiate an indurated deposit, as this may portend only a desiccated soil rather than a rock.

A ripe laterite, which will indurate on desiccation, is sometimes amenable to hand excavation and might be confused with a lateritic soil if the soaking and drying process is not attempted. When soaking and drying these easily excavated materials, it is necessary to prevent the sample from disintegrating during the initial soaking. This may be accomplished by securing the core or cut sample of the material in a tightly fitting burlap bag or cover before subjecting it to the first soaking. If the material is found to be suitably indurated upon the second or third soaking, it may be judged to possess the constituents for induration on drying and is suitable for construction requiring these properties.

Materials which must begin as rock (building stone, riprap, etc.) should be tested by soaking a cut sample for at least 24 hours. When soaked, the sample should be subjected to appropriate striking forces (repeated blows with a hammer) or abrasion (as in the Los Angeles abrasion device) to evaluate the property of hardness. The engineer must decide whether or not the soaked material has sufficient strength to resist the forces to which it will be subjected. The slightest amount of softening would disquality materials sought for use as building stone or riprap. However, a subgrade which employs select, sized, and placed stones could appropriately employ materials in which induration is not complete, provided it is judged by the engineer that the subgrade would be properly drained and would, during its early life, be subject to several cycles of wetting and drying.

There is not presently available a field method of identification of the amount of iron, aluminum, and silica contained in a lateritic deposit. Field observation is the most beneficial means of recognizing materials which possess the properties for induration. A red to reddish-brown predomination in the mottled color of the deposit would be promising, while weak, pinkish coloring on a preponderance of light or yellowish hues would be discouraging. Nearby iron staining draining *away* from the deposit should indicate iron escaping the deposit rather than remaining and this would direct the observer to the destination of the iron rather than its source.

Appendix A refers to a method of differentiating between a laterite and a lateritic soil by means of the Modified AASHO* Compaction Test. Other studies suggest that in the California Bearing Ratio Test, the highest values for laterites are obtained when the sample is soaked at or near optimum

* American Association of State Highway Officials.

moisture content. For materials not indurated, the highest CBR values are obtained from materials soaked at slightly beyond optimum. Materials possess CBR values in direct relationship to their degree of lateritization with a typical value for a well-lateritized material being in excess of 50.

Chapter 4

USES OF LATERITE

EVALUATING LOCAL USES

From the magnificent red and yellow potmarked temples of Cambodia to the durable smooth roads of Gabon, laterite is a stone of a thousand uses. Many are well known to the modern engineer—yet many still lie buried beneath green vines, shifted sands, or cracked stucco.

The imaginative and inquisitive engineer will search a bit further than merely a study of local practices of the particular or related facilities for which he is commissioned. Laterite has all of the good qualities of the splendid porous limestones of Florida which have been used for road surfacing, structural fills and lasting structures such as the Spanish Castile de San Marcus in St. Augustine, Florida. Yet, unlike porous limestone or coquina, lateritic materials must be irreversibly indurated before they will serve as resistant building material. Often, knowledge of their successful use has been lost and must be resurrected. In laterite areas where a high level of culture once prevailed, ruins often disclose laterite used as a building stone. Open cisterns, sewers, headwalls, culverts, flagstones, quays, moles, and breakwaters of laterite have functioned successfully for hundreds of years. Of more recent vintage are private constructions which perhaps were not bound by local tradition.

An examination of historic materials by the geologist will possibly suggest the borrow source, which may be rediscovered after a search. Peripheral and basement laterite may be present in an abandoned quarry. The discovery of a quarry warrents efforts to evaluate past uses and future potential.

BUILDING MATERIAL

When used as a building material, laterite is most desirable in the form of well-joined, small, globular cuirasses. In this form, it is most readily quarried in convenient predetermined shapes. Stone construction is hardly the

Fig. 10. Laterite wall in Libreville, Gabon.

Fig. 11. Thai patio of hewn blocks of laterite.

vogue today having been displaced by rapidly rising concrete structures. Laterite as a building stone will be limited to facing, small gravity structures such as retaining walls or simple dams, and single course block construction such as headwalls, culverts, canals, or stone-surfaced courtyards. The Gabonese wall shown in Fig. 10 or the Thai patio in Fig. 11 are two examples of engineering and architectural uses of laterite. Its lovely textured quality and quiet warm hues make it a most attractive material in either the rain forest or the desert. Within the next decade, we may hope that an imaginative architect will design an American embassy of laterite.

ROAD AND AIRFIELD SUBGRADE

The road and airfield, because they are extensive rather than intensive, usually require both cut and fill. The road and airfield engineer strives to effect similarity of performance within cut and fill sections of construction.

This wish is sometimes obtained by determining the ultimate behavior of cut subgrade and designing fill subgrades to behave in a similar manner. In rain forests cuts may range from mature laterite, to partially lateritized materials, to materials which would never be subject to lateritization. The variety of eventual performances of these subgrades influences, to large measure, the determination of the required performance of filled areas. The materials which underlie potential cuts directly affect the ultimate performance of a cut surface. Consolidation characteristics of the embankment must be understood in order to select carefully the grade line of embankment which, subsequent to settlement, must coincide with adjacent cut surfaces.

Once a final grade for an airfield or road has been selected, the consolidation characteristics of cut areas must be evaluated to determine where the surface of the cut will eventually stabilize. In decomposed igneous or metamorphic rock the removal of overburden will often cause the underlying soils to expand upward, achieving a new level. Subsequent placement of base course and wearing materials, together with the applied traffic load, may reconsolidate the subgrade. It is essential that the performance of cut areas be carefully studied and thoroughly understood. The soils engineer should be familiar with the characteristics of the soils and rock materials which cuts traverse, and having determined upon the ultimate performance of cut areas, he must carefully plan desired performance of fill area to achieve uniformity of transition of grade line from cut to fill.

Having studied the consolidation characteristics of usable natural materials which underlie the filled areas, the soils engineer should evaluate the consolidation characteristics of varying fill materials which might possibly be used for embankments, determine the probable amounts of settlement which natural materials underlying embankments and the embankment materials themselves will undergo, and determine from these evaluations if the materials are compatible with the settlement characteristics of adjacent cut areas.

Inthe event highly compressible soils are found to underlie potential fill areas, they must be removed and replaced with materials which will give a more desirable performance. Ground-moisture characteristics of natural subgrade materials are highly significant in determining soil usage in rain forests. *The most significant cause of subgrade failure in the rain forest is inordinate and undesirable moisture characteristics in subgrade materials.* The forest road or airfield should always be designed to assure a total and continuously well-drained subgrade. However, in assuring proper drainage not only is the sizing of drainage structures which underlie fill areas important, but it is also necessary that the forest vegetation be kept clear of the immediate environs of the construction. Often continuous clearing and grubbing programs must be mounted to assure that vegetation does not encroach closer than several hundred feet to a forest construction. The engineer should plan for clearing the forest a minimum of 100 ft adjacent to construction—200 ft is preferable. If possible, this area should be sloped away from the construction itself to insure that the subgrade is continuously drained. Sometimes these provisions and the attendent costs have been inadvertently omitted in preliminary cost studies or in final design plans. These steps of clearing the forest away and keeping the subgrade drained are more essential than proper surfacing.

In selecting materials for use as subgrade fill, the sensitivity of lateritized, but nonlaterite, materials should be carefully studied to assure that they will not be given to adverse performance as a result of future soil-moisture variation during the life of the project. Some lateritic soils are moisture sensitive and lose strength with an increase in moisture content. The sensitivity of an embankment material may be discovered by preparing compacted samples of the materials at the lowest density appropriate to the construction and subjecting these samples to strength tests at varying degrees of moisture up to total saturation. The minimum strength of the soil is appropriate for design use. The sensitivity (loss of strength through moisture increase) should be the measure for determining the acceptability of the soil for the construction. The engineer should beware of a soil which loses more than 25% of its strength upon soaking.

Laterite materials are normally too scarce for use in subgrade fills. Most likely subgrade fills will be constructed of lateritic soils. The soils engineer should take special note of the moisture content of embankment materials during both the dry and wet seasons and relate these to the moisture contents which are necessary for minimum compaction of the design subgrade. If it is found that an embankment material has a moisture content which is in excess of the moisture content at which a desired degree of compaction can be achieved, it must be recognized that aeration will be necessary before the moisture content of embankment materials can be reduced sufficiently so that compaction can be achieved.

This presents the problems of construction seasons, the possibility of reducing soil-moisture content through aeration, and the need of progress during a work season. If, during the dry season, embankment materials possess moisture contents which are in excess of that from which proper compaction may be achieved, and if climatic conditions are such as to render the successful aeration of soil highly unlikely, then other means must be found to obtain proper soil strength in subgrade materials.

A common and desirable solution is to plan to use embankment materials at natural moisture contents and to compact the soil to the maximum density achievable. In the event sufficient soil-embankment strength can not be achieved, strength additives must be found which will overcome the deficiencies, or they must be overcome by increasing the thickness and/or strength of base and wearing courses.

In summary, the soils engineer building in a laterite area must recognize that he will normally deal with highly sensitive lateritic soil, rather than insensitive soil or laterite. These materials will be found in borrow areas at moisture contents which are close to, if not in excess of, the moisture contents at which desirable degrees of compaction can be achieved. Yet practically, because of the high moisture content in air and frequent forest rains, drying cannot be accomplished. Subgrades must be designed within these parameters. Otherwise the construction cannot be performed in a reasonable period of time.

An interesting use for tropical soils as subgrade was noted by the author in the design of a road in West Cameroon. The results of this design, including the procedures necessary to evaluate the construction materials and the steps necessary to the final selection of degrees of subgrade compaction and thickness of base course and wearing surfaces, are presented in the following exercise.

This road dictated a minimum budget which required the use of minimum thickness of stabilized subgrade and wearing surface, necessitating maximum strength and reliability of subgrade behavior. The maximum moisture content of the earth materials in this, as in other tropical zones, was beyond the upper limit of workability in order to achieve high strength through soil compaction: but aeration of soils in order to reduce the moisture content was not indicated due to the short construction season and the remoteness of construction areas which precluded the use of sophisticated construction equipment.

The direct relationship between the modulus of subgrade reaction, the thickness of the base course and wearing surface, and the intensity and dimension of wheel loads are well known. It is necessary to achieve the optimum cost balance between the three factors—cost of preparation and compaction of subgrade, cost of base course and wearing surface, and the maximum wheel load required by the design. Since wheel loads could not be varied in design consideration, only the quality and quantity of base course and surface materials and the strength and deflection characteristics of the subgrade soil could be changed.

The soil strength at this location is a function of soil density and moisture

Table 6. Laboratory Test Results

Typical Soil Characteristics

				Modified AASHO		
Station	Soil type	Field moisture, %	Field density, lb/ft³	Maximum density, lb/ft³	Optimum moisture, %	Laboratory CBR
12+245	SM	23.3	91.2	121.5	14.5	11
77+260	SM	13.9	107.2	121.0	12.0	28
4+455	SM	31.4	85.6	104.5	16.0	39
30+00	SM	39.7	67.6	102.0	17.0	29
49+00	SM	23.6	91.4	110.0	19.0	37
109+25	SM	36.0	72.3	98.0	24.0	22
132+200	SM	30.0	93.0	97.0	25.5	32
399+150	SM	13.6	102.2	120.0	9.0	29
107+849	SM	29.3	94.4	98.0	23.0	25
430+119	ML	32.8	66.3	98.5	21.5	27
347+000	SC	14.4	123.2	125.5	13.6	38
245+050	SM-SC	19.5	110.9	113.0	16.0	52

Strength Test Data

Station	Soil type	Field moisture, %	Field density, lb/ft³	Unconfined compressive strength, lb/ft³
19+12	SM	16.5	117.5	4.940
13+250	SM	29.1	86.9	9.620
101+000	SM	20.3	71.8	4.230
134+200	SM	32.4	78.5	7.190
166+050	SM	42.0	74.7	7.190
215+150	SM	18.5	109.7	4.530
228+60	SM	35.0	82.3	6.970
245+000	SM	24.6	100.5	4.320
433+448	SM	15.0	106.4	24.210
413—"A"	SM	7.3	86.4	4.010

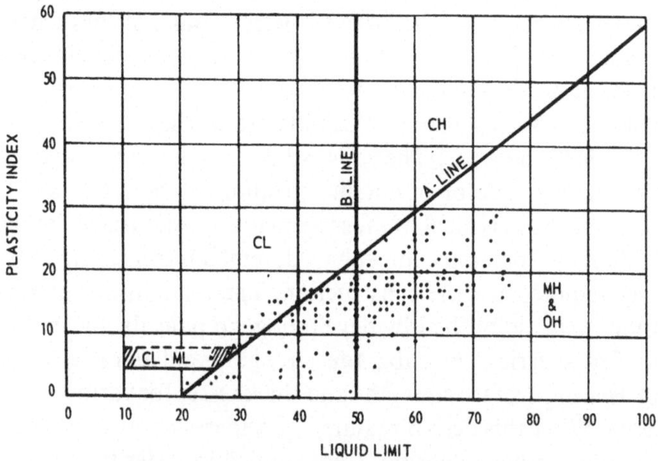

Fig. 12. Plasticity chart, Atterberg limits.

content. The moisture–density relationship for a given compactive effort illustrates the critical nature of soil moisture to density and strength.

The characteristics of the subgrade soil materials were obtained by conducting an extensive soil exploration along the route of the proposed construction and testing selected samples in the laboratory to determine certain

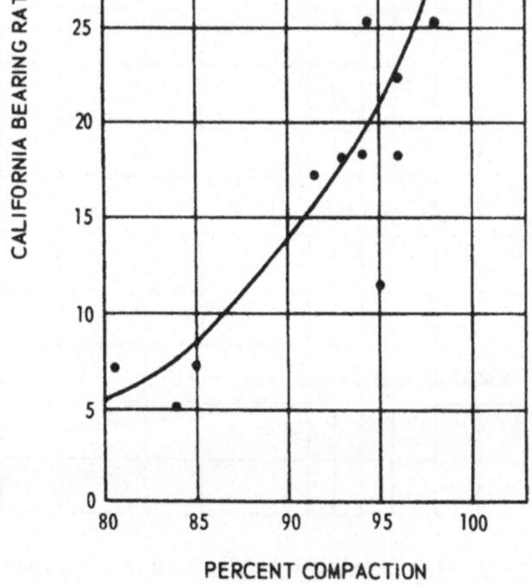

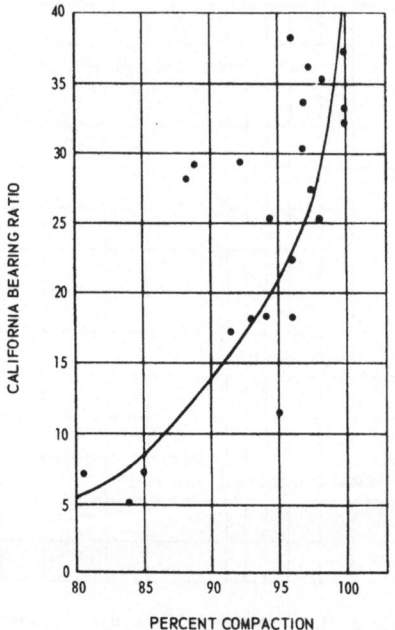

Fig. 13. Compaction *vs* CBR relationship.

of their physical properties. The laboratory test results of these materials are given in Table 6 and shown in Figs. 12 and 13. Two conclusions can be reached from close examination of these data. The appreciable variation of the California Bearing Ratio values illustrates the variability of the soil materials and field moisture. The soils, while less than saturated, were considerably to the wet side of optimum moisture content. The soil materials analyzed in Table 6 are typical of those available for natural subgrade and as sources of borrow materials for subgrade embankment. An H–20 design loading was required for the construction. The relationship of road-design considerations was established by developing graphically the minimum requirements for a variety of subgrade strengths and thicknesses and base-course and wearing-surface strengths and thicknesses for various intensities of wheel loading. While this design required consideration of only one intensity of wheel loading, other intensities were studied in order to see the relationship of varying loading. The results of a typical plot of strength of subgrade *vs* base-course and wearing-surface thickness are shown in Fig. 14. Based on the parameters established by this analysis the general minimum strength requirements for the soil were defined. Compaction tests were performed to determine the degrees of compaction required to achieve the required strength

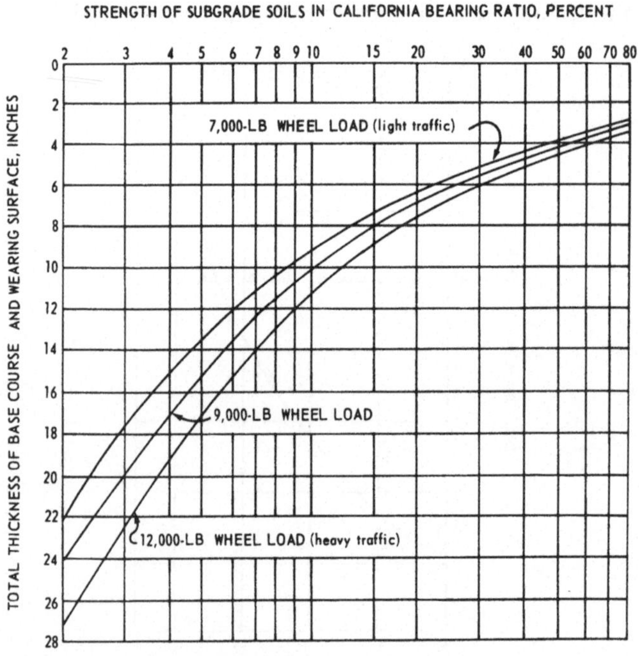

Fig. 14. Strength of subgrade (for total depth of wheel load influence) *vs* base-course and wearing-surface thickness.

parameters. A plot of this relationship is shown in Fig. 13. It will be noted from this curve that a minimum compaction of 90% of maximum density was necessary in order to achieve CBR values of 15, while 95% of maximum density was necessary to achieve CBR values of 21. An examination of the compaction test characteristics of the soil and the comparison of these with natural moisture content revealed that field moisture content precluded achieving 90% compaction in 55% of the soil specimens examined. In 77% of the soil specimens examined 95% compaction was precluded. The soil specimens were obtained during the period of September through January after the end of the rainy season. The conclusion reached through an examination of these data was that moisture conditioning would be required for embankment soils. The difficulty of moisture conditioning under anticipated construction environment strongly indicated the desirability of reducing soil strength requirements of the embankment soils.

It was then necessary to determine the relationship between the costs of various thicknesses of base-course materials and the costs of achieving various degrees of compaction and corresponding soil strength. The results of these analyses are shown in Figs. 15 and 16. While there is no accepted means of evaluating the difference in cost between the various degrees of moisture conditioning and compaction, the experienced designer may assign appropriate values for differences.

In the southernmost two-thirds of the route traversed, soils were of lesser strength than in the northernmost two-thirds. *Natural* soils strengths at grade exhibited CBR equivalent strengths of about 10 to 15 in the southern-

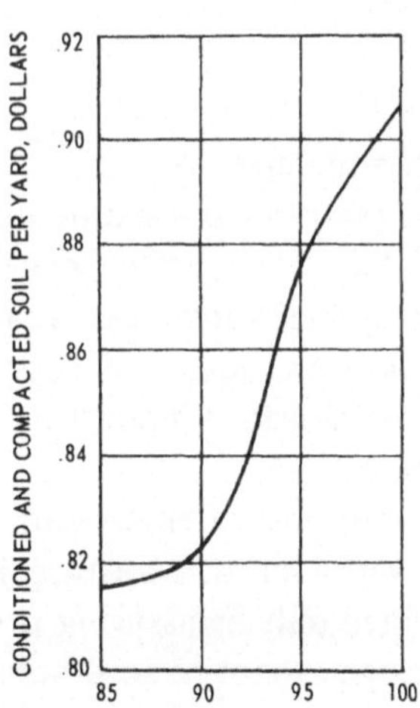

Fig. 15. Cost relationship for soil compaction.

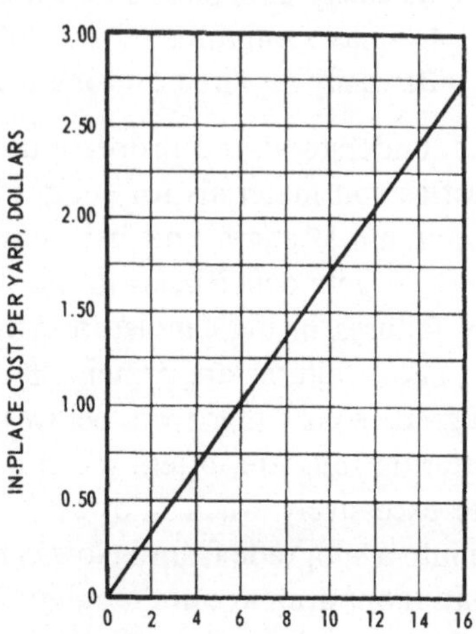

Fig. 16. Cost relationship for gravel base.

most third and of 15 to 25 in the northernmost two-thirds of the route. From an evaluation of these data, it was decided that in the cut areas of the southernmost third of the road, the top 12 in. should be compacted to 95% of the maximum dry density, while in cut areas of the northernmost two-thirds, the top 6 in. should be compacted to 90% of the maximum dry density. All fill was to be compacted to 90% of the dry density. A crushed-stone base or compacted laterite was designed to be placed at a finished thickness of 8 in. on the entire route. A wearing thickness of stone or crushed laterite 1 in. thick was designed.

While a wearing surface of blacktop characteristics was desired for this construction, it was the designer's belief that either laterite surfacing or a fine crushed-stone surfacing would be more appropriate, as these materials can lend themselves to more attentive maintenance practices and would not lull maintenance personnel into a false sense of security.

BASE COURSES AND WEARING SURFACES

The excellent laterite-based and -surfaced roads of Libreville, Gabon, and its environs attest to the value of the material for this use. The numerous reported failures of laterite roads elsewhere in West Africa and in southeast Asia simply confirm its misuse.

To be successful, a base and surfacing and the material from which it is constructed must:

1. Be readily identifiable and available.
2. Be easily understood and used.
3. Display symptoms of impending failure.
4. Be easily repaired and maintained.

Laterite, understood and properly used, is such a material.

Just as soil materials for subgrade must be realistically used to obtain their optimum strength and behavior, so must base and surface materials be used to negate deficiencies in the subgrade. It is well known that road or runway failures in high moisture areas are usually the result of subgrade failure. These failures are principally the result of moisture infiltration into the subgrade soils either from below or from above.

Lateritic soils are, often, moisture sensitive and lose strength and consolidate excessively when allowed to become waterlogged. Moisture infiltration into a subgrade from below can be prevented only by assuring proper and continuous drainage of the subgrade. This design cannot be accomplished by employing customary methods applicable to low moisture areas and insensitive soils. Every reasonable effort must be made to prevent ground and surface water from saturating subgrade soils either cut or fill.

Procedures for carefully sizing drainage structures to prevent blockage of surface drainage are reiterated. The installation of drainage blankets below subgrade fills is necessary if a positive head can be expected to force water into the fill from below or if subsurface drainage within the fill would be impeded without such a blanket. Clearing of forest cover adjacent to a construction and terrain sculpturing to assure fast dispatch of surface water are essential. Even with these precautions, freshets or the downpour of the rainy season may accumulate sufficient runoff to cause ponding on the uphill side of an embankment. This will, at least temporarily, inundate the embankment.

Next in design are the problems of subgrade behavior, proper behavior of the base and wearing surfaces, and moisture infiltration from above. The deformation and strength characteristics of a subgrade soil may be realistically obtained by the CBR test and by the Unconfined and Triaxial Compression Tests.

With a knowledge of the strength and deformation characteristics of the subgrade, and the wheel loads and traffic frequency of the design requirement, the engineer selects the optimum base-course and wearing-surface thicknesses to support applied loads and achieve the desired performance.

The wearing surface must receive wheel loads without immediate failure and the base and wearing surfaces must remain impermeable. Otherwise, moisture from above will infiltrate the subgrade causing it to fail. This in turn will cause lack of support at the surface causing further failure by repetition of the cycle: moisture infiltration, subgrade failures, surface failures, and further moisture infiltration and further subgrade failure.

Recognizing loss of subgrade strength and increase in compressibility due to increase in soil moisture in lateritic soils, the engineer should use them with a slightly higher safety factor than would be employed with an insensitive soil. This increased factor of safety should be translated to an increased thickness of impermeable base and wearing surface.

A typical design plot of the relationship between strength of subgrade vs base-course and wearing-surface thickness for light to heavy wheel loads of road traffic is shown in Fig. 14. Similar design data may be developed for aircraft wheel loads. The criteria of this design presume base-course and wearing-surface strength and flexure characteristics which are associated with a laterite possessing CBR values of 50 or greater. When properly placed, crushed, moistened, and compacted, laterite will easily achieve such requirements.

In determining material requirements and evaluating shrinkage through crushing and compaction, the materials should be studied by the use of test sections constructed from laterite secured from potential borrow pits. Such tests should be preceded by laboratory tests, compaction and CBR, in order to learn crushing characteristics and optimum moisture and strength values.

As a general guide, the thickness requirements in Fig. 14 of base and wearing surfaces should not be reduced for base-course CBR values of less than 80. For values in excess of 80, the total required thickness may be reduced by 27%.

Laterite from borrow pits frequently arrives at location in large chunks and must be crushed before spreading to uniform thickness. A grizzly or heavy wire-mesh roller is suited to this purpose as it crushes to workable size without actually densifying. The Engineer School in Appendix A, however, warns against unnecessary structural degradation. We recognize that their admonition is directed against loss of strength due to too fine a gradation. As a guide, the engineer should consider the size and distribution of crushed particles and keep these within limits that will allow a dense mass when compacted. For proper compaction as a dense base course, laterite should be no greater than 3 in. in maximum size. Once this maximum has been achieved, preparation and compaction may begin. Pieces which resist crushing by the grizzly should be broken by hand or removed from the blanket. As with other friable rock materials, the rock should be compacted at, or slightly above, optimum moisture.

The prepared subgrade surface should be sloped and crowned to assure proper surface drainage during preparation of base material for compaction. Laterite loses moisture when exposed. Aeration is not difficult to perform in the event the laterite borrow is at a moisture content in excess of that which is desirable for proper compaction.

Watering in arid climates presents a more difficult problem due to rapid evaporation of surface moisture. Careful scheduling of watering, with immediate compaction, will assure densification and sealing before evaporation.

In selecting compaction equipment for densifying laterite, care must be taken to assure that the equipment is not so heavy as to damage the underlying subgrade or to over degradate the laterite. On the other hand, the equipment must be sufficiently heavy to achieve required density without inordinate rolling or nonuniform densification. An optimum thickness of compactible laterite is usually an uncompacted blanket of no greater than 6 in.

For this thickness of laterite, tandem sheepsfoot rollers weighing 8 tons usually will achieve a density of 95% of the maximum density of the Modified AASHO Test with four to six passes. An 8–10 ton steel-wheeled vibratory roller is effective with about the same effort. Pellet laterite is most easily compacted with smooth-wheeled or pneumatic tire rollers of the 5–10 ton range.

For base courses requiring multiple lifts of compaction, care must be taken to assure that the lower compacted surface is roughened and moistened before applying further uncompacted material. This is quite important as it assures that the base will act in flexure as a single unit, rather than a stack of separate plates. Furthermore, a uniformly compact and properly joined base will properly inhibit moisture infiltration.

The selection of the balance between base-course thickness and that of the wearing surface is usually determined by the smoothness requirements of the construction surface and the sophistication of maintenance which may reasonably be expected during the life of the roadway.

A laterite-surfaced construction, properly maintained, will perform adequately in desert, grassland, or rain forest. Compacted laterite is, in itself, sufficiently impermeable to resist surface water and surface treatment need not be employed to fill this need. Under conditions of wear which smooth a surface, wet laterite surfaces may be less skid resistant than desired. In addition, laterite is more subject to surface wear than asphalt-sealed roads and, at the outset, this may seem to indicate a greater need for maintenance. This need is offset in primitive locations, however, by the casual maintenance practices which are usually attendant to supposedly paved-road surfaces. There is a fallacious belief that once a surface has been blacktopped it will forever be free of maintenance. Surface failures which appear in a blacktop are less expeditiously repaired than those in compacted-rock surfaces. An abraded blacktop surface or pothole can simply be avoided by traffic and is not as appropriate a reminder of the need for maintenance as is the abraded rock surface.

Because of the usual remote location and minimum traffic of laterite construction, wearing surfaces of minimum thicknesses are usually in order. Several inches of surface is sufficient for roads and a corresponding minimum thickness for blacktopped surfaces of light runways or hardstands.

Surfaces of heavy-duty runways, on the other hand, should be thicker than would be the practice in nonlaterite areas. This requirement is necessary to assure an extra margin of surface strength and impermeability, attendant with unsophisticated maintenance practices and the critical nature of landings on heavy-duty runways.

The choice of wearing surfaces for heavy-duty use are dependent not only on the availability or the cost of importing materials but also on the ability of a surface to display symptoms which show need for maintenance and the ease with which maintenance can be achieved without disrupting operations. As a general guide, asphalt surfaces are more amenable to constructions over subgrades of minimum soil strength in which failures might be more likely. However, a better guide is to design adequately and protect subgrades against failure.

EMBANKMENT FOR DAMS

Dams present a special problem for the engineer who would use laterite as a construction material. Rock fill and rock-material dams are usually utilized only at locations where suitable earth materials are scarce and rock materials can be more economically transported and utilized. Some of the rocks suitable for crushing and compaction as part of a dam embankment are sandstones, shales, decomposed granite, gneisses, schists, ore tailings, and some slags. In a compacted state these materials are far more permeable than silty clay or clay soils of low plasticity and cannot normally be employed as impermeable blankets or to form impermeable sections. Laterite, however, has some of the same recementing characteristics as coquina. Therefore it is useful as a material for an impermeable barrier.

Usually, though, laterite deposits are scarce and the cost of moving and placing laterite materials is too great, compared with that of handling other lateritic or nonlateritized soils locally available, rendering laterite economically unsuitable for dam construction. In considering the use of a laterite as a select material for use in particular sections, careful consideration should be given to the permeability characteristics of the material as compared with the variety of materials which are available for the construction. It must be ascertained prior to selection of a laterite material for use in an embankment whether its strength and permeability parameters are reasonably uniform, whether they fall within a narrow range, and whether the material can be placed, crushed, moisture-conditioned, and compacted economically. In order to properly assay permeability of laterites, relatively large-sized permeability tests are indicated. However, these tests cannot be adequately performed at field locations. It is most desirable that appropriate samples of a variety of the laterite materials (bag samples 50 lb in weight) be shipped to a well-equipped laboratory and a battery of permeability and strength tests performed.

Permeability tests for laterite which is to be compacted should be performed on materials with grain size no greater than one-half inch in the largest dimension. This will produce specimens with characteristics comparable to embankment laterites which were crushed before compaction to a size no greater than six inches in the largest dimension.

Laterite materials have singularly high compacted strength, and when employed in embankments for their strength and permeability characteristics are most effective in improving the total strength of the dam profile. As the strength and permeability characteristics of soil materials which must be utilized along with laterite materials for dam embankments are so varied, no generalized conclusions can be reached concerning the proper balance between the laterite, lateritic, and nonlateritic soil materials. Laterite is useful for utilization as a core when crushed and compacted, as dam facing, riprap, and for selected *toe drain* materials, provided in the latter instance the rocks are placed carefully by hand to prevent crushing of individual rock pieces.

In properly designed dam sections, toe-drain rock is covered with a surface blanket of soil so as to prevent the section from being abraded and eroded by surface water. Discharge wingwells through discharge structures can protect toe drains from being abraded by discharge water. However, when laterite is employed as a medium of wave protection on the upstream face of the dam, the wear-resistant characteristics of the rock must be carefully tested and documented. The best size of laterite rocks to use for an upstream facing to resist wave action is the typical "one-man" sized rock, which is the largest rock which can be carried a short distance and placed by a single laborer. To observe the performance of laterite as riprap, a test section of the intended laterite of the sizes intended for use in the dam should be subjected to the violent action of water. This can often be accomplished by making a pile of the rock within a nearby stream during preliminary design studies. The characteristics of these materials to resist movement by water forces as well as abrasion and deterioration should be observed over the longest period of time practical before deciding if the materials are suitable for use as upstream riprap. However, even if the laterite does not have the properties to resist abrasion and deterioration, it may be the only rock available within the area for such a use. In this event, maintenance and replacement of the damaged laterite rocks in riprap sections must be balanced against the cost of individually-cast concrete riprap blocks, or perhaps against the more expensive concrete blanket from the upstream face.

Compacted laterite covered with an impermeable blanket, such as an asphalt coating, may prove more effective and less expensive than a concrete-stabilized soil blanket to cover the dam surface. In order to judge if a particular laterite material has suitable water-resistant characteristics the engineer should observe the behavior of similar laterite used in surface-protection blankets in construction elsewhere.

Laterite is excellent for use as a surface covering in arid areas where the rainfall would not support grass and it could not be employed as a medium of surface protection. In the tropical climate, however, it will frequently be desired that the whole dam section be protected by heavily grassing both the exposed upstream and downstream faces. In this case, it is quite important that the materials selected for the facing not be such as would lateritize in a short period of time, as lateritization effectively inhibits the growth of surface covering. Grass should only be employed on earth materials in which the identifying ratio of silica to sesquioxides (according to the equation of de Medina, Table 1) is in excess of 2.5. It is not anticipated that these materials would become indurated as long as the surface remains moist. This is usually true, since there is a deficiency of feeding aluminum or iron salts which would reduce the ratio. However, it must be remembered that a blanketing material can serve as a feeding medium to underlying soils and that sesquioxides fed through the blanket could be sufficient to cause lateritization of underlying material effectively inhibiting the growth of surface vegetation in the blanket.

Therefore, even though the surface materials have been ascertained to be of material not subject to lateritization, tests should be made on materials immediately underlying to insure that they too, will not lateritize.

When practical, it may be desirable to provide a blanket of strong topsoil which contains humus and will support plant growth. This is usually quite scarce in rain-forest areas and totally lacking in arid regions. Small deposits of organic materials can be usually found in alluvial plains or stream beds and can be effectively introduced into the design. Bar deposits of materials such as coarse sands and gravels can also be used as a surface coating, as they will usually support surface vegetation and are sufficiently inert so as not to provide a feeding medium which would serve to lateritize underlying materials.

If possible, indigenous grasses should be employed as surface covering rather than attempting to introduce new and exotic strains which might bring unusual nitrogen concentrations to bear on the soils and in turn promote or hasten lateritization.

Chapter 5

CONSTRUCTION TECHNIQUES

BORROW DEVELOPMENT

The laterite borrow pit is developed much the same way as a borrow for other materials of a select nature for use in construction. The surface is cleared of all trees and vegetation and of the sparse covering of organic soils which are usually present. Soils overlying the laterite rock are removed so as to expose a fresh surface of rock to be worked. It is usually desirable, as in other borrow areas, to conduct the borrow operation as a side-hill borrow, beginning at the lower extremities of the deposit to be worked and working uphill. In laterite, this has the advantage of providing a well-drained work area and usually affords an increasing thickness of rock exposed on the cut face, as generally, laterite deposits are thicker at the upper extremities than at the edges.

While it may be possible to dislodge and secure laterite by means of normal quarry excavating equipment, the cost of this equipment in a remote location and its maintenance usually render it uneconomical. In the non-porous laterites, skillful blasting can produce material for direct loading. Once a vertical cut face 4–6 ft in height, or greater, has been presented in the borrow area, experimental blasting techniques can be utilized to establish effective means for securing dislodged materials.

In order to have a beginning point in experimenting for a successful blasting procedure, it is suggested that for a laterite cut face 6 ft in height or less, shot holes be drilled a distance back from the cut face equivalent to one-half to two-thirds of the height of the cut face and to the total depth of the cut face. The suggested spacing for these shot holes would be the height of the cut face. The shot holes can be loaded with a half stick of 40% gelatin dynamite placed each 3 ft of hole depth, with intervening holes between charges refilled with crushed and tamped laterite. The charges should be placed no closer to the surface than 3 ft. If available, delayed caps should be used to effect a delayed explosion progressing from one end of the cut face to the other. Simple, shot-hole drilling equipment is suitable for drilling the blastholes. A suggested drill is the lightweight wagon drill powered by a minimum-sized portable air compressor.

Pieces of laterite larger than 1/2-yd size (approximately 1½ ft in greatest dimension) should be reduced in size by breaking before loading.

Often, a front-end loader of the D-6 class or greater, or power shovel greater than a 1-yd class can be effectively employed for dislodging and loading laterite. If such equipment is available, its use would prove a much more suitable method of dislodging laterite than by blasting. An attempt should be made to work cut faces by this method before resorting to blasting. The method selected for dislodging laterite rock should be the one which is proven most expeditious, but also most effective in reducing the size of dislodged pieces so that little additional effort must be used in order to have suitable dislodged material available for transportation and placement on the construction surface without undue extra breaking effort.

Special consideration must be given to dislodging materials where large rock sizes are desired such as in dam riprap or toe drains. When larger sized fragments are desired, it is necessary to reduce the explosive charges so that large rock masses are dislodged and these in turn can be broken into roughly desired shapes before loading on the conveyance. If the construction requires them, it is possible to secure special-sized laterite rock materials in conjunction with normal borrow operations. Usually oversized rocks are dislodged as more resistant pieces of laterite, and consequently are more effective for the intended purpose. If these materials can be secured at the time they are dislodged and stockpiled, the selectivity process of choosing such rock sizes has the advantage of probably assuring the selection of the most desirable and resistant large rocks and also negates the requirement of breaking these rocks into more usable sizes for employment in crushed and compacted blankets.

INDURATION IN AIR

Often tests will disclose areas of lateritic soils which if uncovered and subject to appropriate periods of desiccation would effectively become laterite in several seasons. Once located, preparation requires clearing of the area and proper sculpturing of the exposed soil surface to assure that it will remain well drained and as nearly dry as possible throughout the wet season, prolonging the length of time it will be subject to desiccation.

Sometimes it is more effective to dislodge materials which are in a near laterite state than those materials which are partially indurated but not irreversibly so. Such materials, if placed in a loose pile of large rocks, and is well drained and subject to all around exposure to air drying, may be expected to become indurated in one to two seasons. This method has been observed by the author to be employed effectively in Gabon. In certain areas which are not irreversibly indurated, materials are cut from the borrow and stockpiled within the immediate environs or transported to locations of desired use, carefully placed and stockpiled, and allowed to remain until they have been properly indurated by several seasons of high desiccation. This work, of course, requires careful planning and proper selection of materials, as well as ability, resources, and willingness to stockpile and ripen materials for future use. For the construction of an extensive project such as a large dam, riprap or toe-drain material can usually be discovered, quarried, and indurated during the several seasons between working borrow areas and the final use. However, in this regard, particular care must be taken to assure that the materials will indeed indurate during the interim. A crude but effective test is to take lateritic soil and subject it to repetitive soaking and drying in order to evaluate whether induration is taking place. A crude estimate for the relationship between the air temperature during natural periods of drought and the temperature of artificial drying is shown in Fig. 17. A drying temperature of 150° F for artificial drying is believed to induce no properties which would produce physical or chemical changes unattainable through a natural drying period.

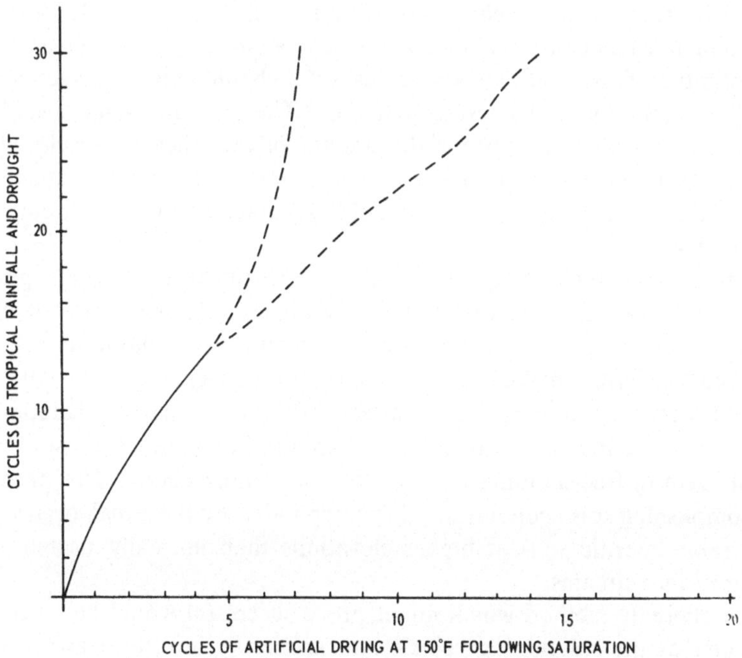

Fig. 17. Relationship between natural and artificial desiccation.

ARTIFICIAL INDURATION

Rarely will the cost of laterite be so expensive as to render artificial induration economical. Many lateritic soils which have not been indurated will, under the proper circumstances, become indurated during several seasons drought, particularly if they are cut from the borrow and properly stockpiled in exposed positions so that the drying process can function properly.

Experiments with artificial induration are of increasing interest to the student of laterite. Suggestions for experiments in artificial induration are explored in this section. In Chapter 1, page 8, chemical action of lateritization as well as kaolinization is described. Chapter 2, page 13, describes some of the factors which are present in the unknown phenomenon of induration. It would appear that lateritic soils not yet ripe (ready for induration by desiccation) must be properly made ripe by reducing the silica content to approximately 16% and increasing the iron content to approximately 45%.

Approaching the simpler problem first, it would appear more economical and easier to reduce the silica content than increase the iron content. Warm alkaline waters are expected to dissolve the silica materials, or possibly reconstitute them as colloids. Naturally occurring alkaline waters are rarely associated with laterite locations; however, waters may be artificially alkalinized rather inexpensively by diverting a small stream over a feeding bed containing limestone or basic minerals. If it is possible to increase the degree of alkalinity in flowing groundwater, this water should then be deflected to a deposit of nonindurated lateritic materials. The alkaline water would be sufficient to dissolve an appropriate amount of the silica materials which inhibit induration. Once the desired silica is removed from the dissolving water, the remaining lateritic soils would be more receptive to induration through drying.

This scheme might be quite difficult to perform. It must be recognized that nonindurated lateritic soils with the grain-size characteristics of clays and silty clays tend to expand and soften, when saturated without appropriate surcharge, rendering ineffective the previously established cuirass structure. It would seem more appropriate to attempt to remove silica by solutioning from *in situ* lateritic soils rather than attempting to remove the soils and present them in free-standing form to the solutioning action. The problem of accomplishing this economically is compounded by the usual occurrence of desirable lateritic soils at higher elevations than normally usable free-flowing surface streams.

An alternate method which might prove successful would be to utilize rain water as the solutioning vehicle and artificially introduce an alkaline medium overlying the soil mantle from which it is desired to reduce the silica content. An experimental method for this latter scheme would involve the location of a lateritic soil deposit high in silica which could be cleared and diked before a rainy season. The area enclosing the dike should be covered with an artificial soil mantle several feet in thickness to which has been introduced a 15–24% concentration of lime or relatively soluble limestone. Several seasons rainfall under such conditions followed by removal of the covering mantle and natural or artificial drying during several droughts might produce usable indurated material.

These thoughts are presented as suggestions for experimental procedures at locations where nonindurated lateritic materials are replete or in an area which has been depleted of laterite and where further laterite is critically needed for maintenance.

Considerable further effort is needed in the study of artificial induration. Heretofore, rock has not been sufficiently scarce to require these drastic and expensive measures. Certainly, there is promise for developing an understanding of the processes of induration and for devising economical means to effect them. *It is only when* the desire for the possession of these materials becomes sufficiently acute that we must resort to such means as the foregoing.

PLACING BORROW ROCK

One of the more interesting techniques for employing laterite in road subgrade is that used in the construction of city streets and arterial highways entering Lagos, Nigeria (Fig. 18). As the area is one principally underlain by recent alluvium, laterite is rather scarce and must be imported. There are,

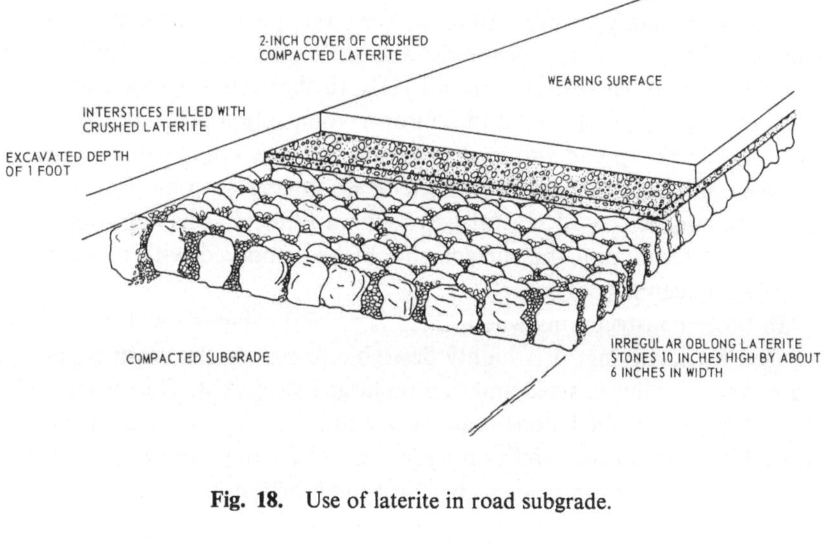

Fig. 18. Use of laterite in road subgrade.

however, good sources of laterite gravel just outside Lagos, extensively used for road construction. Most likely, in order to get maximum use of strength a novel subgrade scheme has evolved which is illustrated in Fig. 18.

Within a prepared base excavation of proper depth, the rocks (6 by 12 in. or thereabouts) are carefully placed upright so that they are securely adjacent to their neighbors. The voids between rocks are then seeded with crushed laterite which is then compacted. Sufficient laterite is seeded to furnish a laterite blanket of 2-in. thickness overlying the previously placed stones. The advantages of placing the individual pieces upright seems to be that it utilizes the greatest available section of rock and minimizes the likelihood of crushing an individual piece under load.

Of course, such a blanket disconnected laterally except for friction between pieces of rock, must rely heavily on the strength of the subgrade to resist individual rocks from being forced into the subgrade. This scheme can be effectively employed where a subgrade of moderately high and *uniform* strength is present or can be achieved.

Normally, laterite is transported to its place of use in the state and size in which it was excavated. If this gradation is unsuitable for compaction, the larger inclusions must be further reduced in size. Laterite is not usually passed through a crusher—probably because in the remote areas where it is most frequently dug the few available crushers are used to crush more durable country rock. Laterite is usually broken to suitable size by hand. This procedure, which results in nonuniform size, tends to make a uniform compacted density more difficult to achieve. Placement of rock becomes quite important if the following phases are to be effective.

Quite likely the rock will be transported in 2½-ton dump trucks as these are an effective size for such conditions. After dumping, the small loads must be spread thoroughly to reveal all rock. The larger pieces must first be broken by hand and this is a time consuming task but is so necessary that it warrants meticulous care. When it is done properly further crushing by mechanical means can be accomplished and uniform compaction achieved. However, large rocks in the spread will spoil crushing and later compaction.

Once all rocks have been reduced to no greater than an 8-in. size, the blanket should be passed several times with a grid roller having 4-in. maximum grids. Thereupon the surface should be smoothed with a blade and compaction begun.

In those constructions where there is no soil subgrade which might be damaged by scarifying, it is highly desirable to scarify the crushed rock to assure that it is all well sized and that no large pieces exist. This process also assures that the crushed-stone blanket is well joined to the material directly below. Should moisture conditioning (either moistening or drying) be necessary this is an opportune time to perform the function.

CRUSHING

In order to effect the strongest and most impermeable covering blanket, it is necessary that laterite in road or airfield construction be transformed from its deposited state to the crushed state most amenable to compaction. In the normal course of engineered construction, it is presumed that a surfacing material will be transmitted from the borrow area to the site of placement and compaction in a state allowing it to be readily conditioned (crushed and moisture conditioned) and compacted to its required behavior characteristics. It is well known that hard-rock base-course material cannot be compacted on a subgrade utilizing the normal crushing equipment available in construction and, further, that attempts to crush such materials once spread will only result in damage to the subgrade by forcing angular cobbles and boulders into the subgrade. As laterite is thought of as a material not as hard or difficult to crush as hard stone, it might be presumed that in-place crushing would be appropriate and could be accomplished in conjunction with compaction.

Normally, this is the case as crushed laterite chunks which can be overridden by appropriately heavy crushing equipment can usually be reduced to sizes appropriate for compaction. Also, sufficient fines are often produced in the crushing activity to render an appropriately satisfactory grain-size distribution. However, as in-place crushing is a function of the crushing force required to break the rock and the resistance of the subgrade, consideration must be given to the most effective and economical means of achieving the required grain-size distribution from the transported mass. Specifications currently in vogue in various parts of Africa and South America (Brazil, Ivory Coast, Gambia, Liberia, Senegal, and Nigeria) provide that prior to compaction the road or airfield's base-course laterite materials be reduced to a grain-size distribution within the limits shown on the grain-size distribution chart in Fig. 19.

It will be noted that this range reflects a bulge between 1 mm and 0.25 mm, indicating an absence of this fraction. Grain-size distribution requirements of compacted base-course materials result from numerous, previous laboratory studies, as well as those performed for the particular construction. Furthermore, the grain-size distribution requirements and specifications reflect an awareness of successful previous constructions utilizing similar grain-size distribution. In planning a specification for a particular laterite to be used in subgrade construction, the designer would be foolhardy to ignore previously established grain-size requirements, but he must recognize the attendant difficulties in achieving these grain-size distributions from an imported material and evaluate the costs for producing such distribution.

It is generally thought that maximum density and wearing characteristics

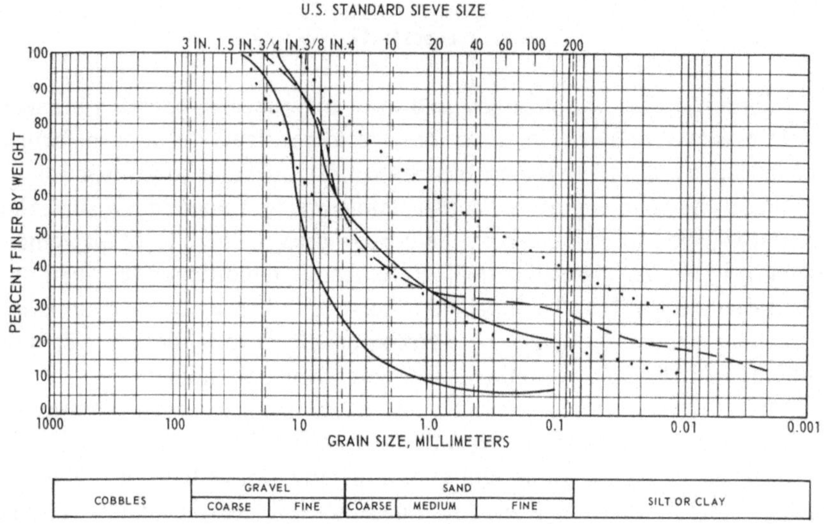

Fig. 19. Grain-size distribution chart. — Range of Ivory Coast laterite, - - Senegalese laterite, ··· range recommended by A. Renillon. (As reported by de Medina.[(3)])

and minimum permeability may be achieved by compacted laterite material which prior to compaction had a grain-size distribution similar to those shown in Fig. 19. It is not, however, practical to attempt to crush such materials prior to transporting them to the construction site. Fortunately, laterite materials lend themselves to crushing when spread on a subgrade of moderate strength by using crushing equipment which does not damage the subgrade by forcing large fragments of the imported material into the subgrade. It is well known and accepted that the material which has a grain-size distribution falling within the limits previously cited will be most amenable to compaction. CBR strength characteristics required for a particular subgrade strength can be most readily achieved with a grain-size distribution falling within these limits. The characteristic bulge in the curves, indicating a deficiency in the sand constituent, may reflect a need for adding a moderate amount of sand material to the mix. The CBR values and the attendant grain-size distribution accompanying the CBR test should be compared with the grain-size distribution obtained in a field test wherein rocky laterite blocks are spread on a prepared subgrade and crushed and compacted using normal field compaction equipment (6–8-ton sheepsfoot rollers or 12-ton flat-wheeled rollers). If the grain-size distribution is found to be corresponding between CBR tests and field tests, it is presumed that the material is amenable to in-place crushing. If, however, field tests reveal a large particle distribution limiting the development of a well-graded, suitably compacted, dense blanket, some means must be found to overcome the deficiency or admixtures must be sought which will result in an artificial realization of the required strength and permeability characteristics.

If it is not possible to achieve the required grain-size distribution by field crushing, the designer must pass the material through a crusher, add a particular fraction to the material, or resort to the use of an admixture such as cement or lime. In Mozambique the following specifications have been adopted for laterite bases and surface materials: for road bases, CBR values greater than 67, a plasticity index of less than 12, and a grain-size distribution in which less than 25% passes the #200 sieve; for surface layers without bituminous protection, CBR values greater than 80, a plasticity index of less than 15, and a grain-size distribution in which less than 35% passes the #200 sieve. The Ivory Coast specifications require a plasticity index of less than 12, less than 25% passing the #200 sieve, and CBR values of 60 for 95% of the maximum density, which is the equivalent of 90 for 100% of the maximum density. Remillon recommends a plasticity index of 15 to 25, implying that the percent passing the #200 sieve should not be greater than 25.

COMPACTING

Compaction should be achieved with the least possible effort, in the mini-

Fig. 20. Smooth-wheeled roller in Cameroon.

mum time, and without damage to the soil which supports the materials that are being compacted. All of these requirements direct themselves to the selection of compaction equipment with particular reference to its weight, use location, etc.

Certainly in West Africa the smooth-wheeled roller, as shown in Fig. 20, is the most popular machine for compacting laterite and volcanic materials in engineered construction. The use of this machine is an outgrowth of the British and French practice on the continent of fixing and compacting a single lift of surface blanket. It is ideally suited to these purposes and if but a 6-in. thickness of laterite is to be used, this machine, of proper weight, has no superior. But since it has the effect of producing a smooth upper surface in the compacted medium, it is hardly desirable if multiple compacted lifts are to be placed, since the smooth surface must be scarified before additional uncompacted material can be placed upon the surface. Otherwise a plane of weakness will exist along the smooth surface. This presence of weakened planes is not presumed in design analysis. Overload tests in base and wearing courses as well as multiple-lift compacted embankments show that horizontal shear occurs along the weakened junctures between lifts. Inherent weakness must be overcome by making this plane as strong as possible.

The sheepsfoot or grill-type roller leaves a highly irregular surface on a compacted lift which is well suited to receive the next lift of material.

Five to eight passes with a sheepsfoot roller should certainly be sufficient to compact an 8- to 10-in. thickness of uncompacted laterite or similar material to 90% of the maximum density achieved by the Modified AASHO method of compaction. If this is not possible, the roller is not of sufficient weight. As a key to selection of a roller of maximum weight for a particular subgrade, the construction engineer may be guided by the rule of thumb that a satisfactory roller will, when empty, compact a subgrade soil of proper moisture content to 90% of Modified AASHO density with eight to ten passes. This same roller, half filled with water, will provide suitable compaction to the first lift of base-course material. Subsequent lifts of base may be compacted more easily with the roller completely filled.

The motor patrol is a useful machine for spreading and windrowing placed laterite, and the scarifying attachments are particularly suited to roughening the surface at a compacted laterite lift in the event it has been previously smoothed with a smooth-wheeled roller. Other scarifying devices, while effective, cannot be used with the care and precision that can be gained with the motor patrol. This care is quite necessary since in attempting to scarify a surface, an inexperienced operator using a machine with too great of a biting power can easily peel off a thin lift of highly densified laterite, particularly the initial lift which overlies the subgrade.

TESTING PLACED MATERIALS

Laterite being considered for engineered construction can be identified by textural examination. Materials which are obviously indurated, such as that shown in the frontispiece (page xii), require but superficial knowledge to be easily identified as laterite. While materials possessing these characteristics will be varied in chemical composition according to the three main constituents, silica, iron, and aluminum, the actual chemical make-up is not of as great an importance as the characteristics of behavior, and these may be more readily determined in the field than by chemical analysis. If for some particular reason it is necessary to determine chemical quality, it is recommended that samples of materials be shipped to an assay laboratory for accurate determination, rather than attempting crude field tests.

The author has observed several very crude field determinations which ostensively resulted in a prediction of the quantity of iron and aluminum salts contained in a rock specimen. Hach Chemical Company of Ames, Iowa, manufactures a field test kit which can be *adapted* to testing the Fe and Al constituents in a solution, and this can be related to the amount of material in a prepared rock sample. The difficulty, it would seem, is getting the metallic compounds into solution. The author has found no readily available means for doing this. At this time laterite which must be analyzed should be directed to an assay laboratory.

The compaction characteristics of laterite materials (both soil and rock) may be determined by a simple compaction test. The Modified AASHO Test (ASTM E-1557-66-T and AASHO T180-61-C) is appropriate for determining maximum density under the compactive effort which most closely resembles strenuous and diligent construction endeavor. Compaction testing reveals the optimum moisture content. For a uniform compaction effort the resulting density increases as the moisture content is increased up to optimum and thereafter decreases as the moisture content increases.

This test requires that all the samples of material should be ground up so as to pass through a #4 sieve, and then a portion of the sample is compacted at a specific moisture content in 5 equal layers in a thick-walled metal cylinder, having a volume of 1/30 ft^3, using 25 blows of a 10-lb hammer falling 18 in. to compact each layer. The wet density of a compacted sample is determined by weighing a known volume of soil, the moisture content by measuring the loss of weight of a portion of the sample when the soil is oven dried, and the dry density from the wet density and moisture content. A series of such compactions is performed at increased moisture contents until a sufficient number of points defining the moisture–density relationship have been obtained to permit the plotting of a compaction curve. The maximum dry

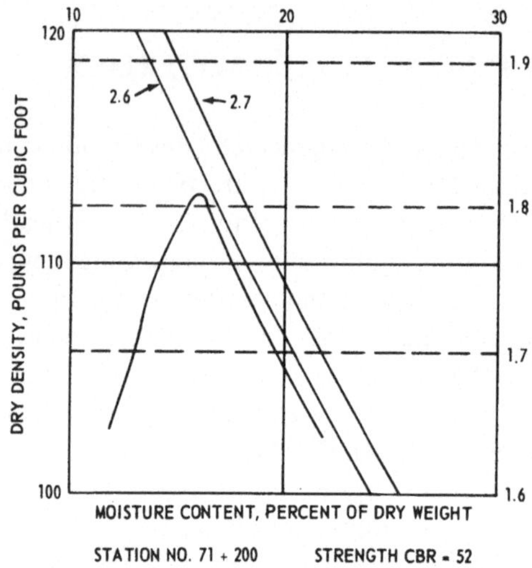

Fig. 21. Moisture–density relationship for a typical laterite by the modified AASHO method of compaction.

density and optimum moisture content for a particular compactive effort are determined from the compaction curve. A typical compaction curve derived from the compaction of crushed laterite is shown in Fig. 21.

The strength of a compacted material may be determined by the CBR test. In this test, the crushed material is compacted at optimum moisture content. Subsequent to the compaction, a surcharge weight is placed on the soil surface and the mold in which the soil was compacted is soaked. After the soaking process has been completed, the mold is placed in a testing machine and a 3-in^2 circular plunger is forced 1/10 in. into the soil. The resistance encountered for the penetration is recorded. The ratio of that resistance to the resistance of crushed rock to the same penetration is taken to represent the CBR value. A CBR value of 100 indicates excellent material with supporting properties as good as those of the crushed rock, while lower CBR values indicate materials of lesser strength and more deflective characteristics. It is from data such as these that the design relationships for strength of base-course materials are derived.

The modulus of subgrade reaction is merely the slope of the stress–strain curve (within the elastic range) for a soil or rocky material. It is normally expressed in pounds per square inch per inch, and is useful in assaying the reaction of subgrade to a wheel load providing the load produces a reaction within the elastic range of the subgrade material.

The field moisture content of construction materials must be obtained in a timely manner so that the knowledge of the moisture content is immediately available. The time-honored method of determining the moisture content by oven drying overnight, while necessary for a check of other methods, is simply too time consuming to be employed on a construction job. It is recommended that some other method be found for immediately determining moisture content at the time natural or borrow soils are being densified. The Speedy Moisture Meter is a very simple method for determining moisture content of construction soils, rock, etc. This instrument, which is manufactured by Thomas Ashworth of Burnley, England, and handled in the U.S. by the Alphalux Company of Philadelphia, Pennsylvania, uses the principle of measuring the gaseous pressure produced when the moisture in a soil sample is combined with carbide powder. A measured quantity (usually 6 g) of a material is placed in a pressure vessel, a measured amount of carbide powder is placed in the vessel's cap, and it is then hermetically sealed and violently shaken so that the reagent and the soil moisture in the soil combine, producing gas. The increased pressure within the vessel is recorded on a pressure meter, which directly converts the recorded pressure to moisture content. Of course, the moisture content obtained by such a method must be checked in the laboratory utilizing oven-drying procedures. However, moisture contents so recorded in the field are found to be quite accurate. Furthermore, they are immediately obtained at the time of need.

In-place density of soil and rock materials compacted in engineered construction must be ascertained by field density tests. At remote locations and adverse circumstances one of the most suitable tests for ascertaining density is by ASTM Test D 1556, the standard test for determining the density of soil in place by the sand-cone method, determined as follows.

The surface of the location to be tested is leveled and a metal template, used to support the sand cone, is placed on it. A test hole is dug inside the circular hole in the template to a depth of approximately 6–8 in. beneath the surface. All soil dug from the hole is placed in a container to be weighed later. The sand-cone apparatus, with a predetermined volume of sand in the glass receptacle, is placed over the previously dug hole, and sand is allowed to flow into the hole until the hole is filled. By determining the amount of sand which flowed into the hole by weighing before and after, and by applying the calibrated pouring density of the sand-cone apparatus, the volume of the hole is determined. The weight of the material which was removed from the hole can be used to calculate the wet density. Laboratory moisture determinations of the material removed from the hole can be used to calculate the moisture content, and from this the dry density of the material can be determined. This can then be applied as an evaluation of the density of the material relative to that required by specifications.

This method is highly successful and most desirable, providing an ample supply of glass vessels can be maintained for use with the sand-cone apparatus. It has been the author's experience that these vessels are subject to breakage in the field and cannot be easily replaced in remote locations. A simple expedient may be employed, however. An operator determines the volume of the hole utilizing dry sand which he pours into the hole dug in the fill. He ascertains his pouring density by pouring loose sand from an open container into a compaction mold. Since the volume of the mold is known, the weight of the sand in the mold can be determined by the operator through repeated pourings. It is necessary that the mouth of the pouring apparatus be held at the same level over the mold for all attempts—say 3–4 in. above the mold. The same pouring speed and generally the same cross section of dry poured material should be sought by the operator in repeated applications.

A skilled density test operator can, by the sand pouring method, obtain sand densities which can be used for volume determinations that vary less than 1/2% from one application to another. The variation by the sand-cone method is not greatly different from that obtained by the sandpour method.

Density of soil in place may also be determined by ASTM Method D-2167, a test of the soil in place by the rubber balloon method. In this case, a template is placed over a prepared surface and a hole dug in the material to a depth of approximately 6–8 in. Over this is placed a rubber balloon together with a calibrated vessel containing water. The water is allowed to flow downward from the calibrated vessel and fill the balloon, which depresses into the hole. The amount of water which is required to fill the balloon in the hole is recorded immediately as the volume of the hole. This application is quite successful for the more sophisticated field jobs and is particularly appropriate for locations where replacement balloons may be readily obtained should the supply become depleted and where the soils are of such homogeneous characteristics that the sides of the dug hole have few protrusions or indentions which could not be properly configured by the balloon's flexible membrane.

For simple soil fills of the type that most subgrade materials are constructed from, a small, driven, density sampler, 2–3 in. in diameter and approximately 6 in. in length, may be driven into the surface of a smooth lift of fill and a sample extracted therefrom which can be tested for moisture and density in the laboratory. This method can be employed successfully where there is little doubt that the material will be homogeneous and where there are few, if any, rock fragments which would seriously damage the results of the test or be damaging to the sampling apparatus. The method is, however, not suggested for rocky materials and should not be used in testing the field density of laterite. It has, however, proved quite successful for use in lateritic or volcanic nongranular soils used in subgrade fill.

The use of a nuclear moisture–density meter can certainly be justified for extensive fill operations, but because of the uncertainty of keeping such a piece of equipment operating, it cannot be recommended for use in remote locations where its repair and replacement would be difficult should it be the only means available for determining the in-place density of materials for construction. Several such machines are currently on the market. The reader is referred to the Nuclear Density Moisture Meter NU-75, presently marketed by Soiltest, Inc., of Chicago. The moisture meter operates on the principle that dense soils absorb radiation and loose soils reflect it. Gamma rays from a radioactive source are deflected into the soil and after an appropriate length of time (approximately 1 min) the amount of reflected gamma radiation is recorded on a dial which has been calibrated to provide the wet density of a material. The moisture content is determined immediately at that time by measuring the number of returning neutrons and thereby determining the number of hydrogen atoms in the soil.

Generally, it is necessary that the nuclear unit be calibrated for each soil to be tested. It is a production-type instrument and can be used on large and extensive fills, but due to the cost of the instrument and its requirements for operation, it cannot at this time be thought reliable for engineering work at remote locations.

Chapter 6

MAINTAINING LATERITE

SURFACING

The functions of an all-weather bituminous or concrete surface for an engineered construction are varied and must be clearly understood by the design and maintenance engineer. These surfacings are impervious to surface water and protect a base and subgrade against its vertical infiltration. Conversely, such surfaces present an impervious membrane which entraps horizontal or upward moving groundwater and can easily be lifted by nominal hydrostatic forces. These surfacings provide added strength to a base course and distribute applied loads proportionately to their membrane strength as well as furnishing horizontal shearing and frictional strength to the medium they are covering (i.e., the base course). When not properly and permanently bonded to the lower medium, they serve as a high-strength, brittle membrane which, when broken, loses all of its inherent strength and thereafter acts merely as a discontinuous, strengthless, added thickness of the upper surface. Furthermore, the rupture of the upper covering provides passages for the introduction of surface water into the hitherto protected subgrade and base course. When the wearing surface is eroded, the frictional properties of the surface to resist vehicle skid become a function of the newly disclosed base course.

The ability of an all-weather surface to function properly depends upon:

1. The proper relationship between the applied load and the wearing surface, base course, and subgrade acting in unison to resist the load when applied repetitively.
2. The ability of the wearing surface to resist deterioration through impact and frictional abrasion by tires (i.e., load).
3. The ease with which surface failures can be noted by only moderately close surveillance.
4. The damage which can occur during the interim between surface failure and repair.

5. The ease, suitability, and cost of repairs to damaged wearing surfaces.

As laterite constructions are normally confined to remote underdeveloped areas lacking in sophisticated surveillance and maintenance, the use of all-weather surfacing presents a dangerous and seldom warranted sense of construction longevity and permanence. In the case of extensive laterite construction such as a rain-forest road, those who use it and those who maintain it too often believe it has been constructed to last forever—maintenance free. Surface failures, when they occur, are often ignored and costly and sometimes irreparable damage is done to the base and subgrade by surface water and/or applied loads.

The decision of whether or not to apply an all-weather surface must be made with a full knowledge of the function of the surface during its expected life. A thin blacktop surfacing is useless, as it will be quickly destroyed by applied loads. Except for airfield construction, the use of thick bituminous or concrete coverings is not appropriate due to their high cost. Yet a minimum thickness of 4 in. is necessary to properly resist abrasion for several years of vehicle loading. It is sometimes not possible to justify the cost of this thickness of imported covering, but a lesser thickness will be easily abraded in a short time and after a few seasons will be more of a curse than a blessing. Such a curse is shown in Fig. 22.

Fig. 22. Inadequate thickness of surfacing on a well-stabilized subgrade.

ROAD SURVEILLANCE

A decision by the designer to blacktop a road in a rain forest should be made only after careful study, with a full knowledge of the sometimes false and frequently dangerous sense of security that a blacktop surface lends such a road. The first requirement of a properly designed and constructed rain-forest road is that it function under the traffic load without either its impermeable surface failing or moisture penetrating to its subgrade, reducing strength and allowing failure. It has been the author's observation that all too often blacktop roads in the rain forest are not subject to appropriate surveillance. Except for the inordinate effects of moisture beating down on the surface of the road, an all-weather road in the rain forest should perform in the same manner as an all-weather road in any other climate. All too often, however, these roads are constructed at minimum cost. Because of the lack of funds, and the lack of understanding of subgrade behavior and the role of a blacktop surface, the owners fail to appreciate the disastrous effects which are always forthcoming when the blacktop surfacing has been eroded and breaks allow water to enter the subgrade. Too often, there is a tendency to breathe a sigh of relief when a blacktop road has been installed and to forget it as being something that does not require surveillance and constant repair.

When the blacktop fails and ulcers form through the base into the subgrade, traffic simply directs itself around, striking the edges of the ulcer, or, quite often, drives on the shoulders. This diverting of traffic causes other failures to occur because loads are being introduced at locations which were not intended to receive them. The pattern then becomes progressive and soon the entire road surface is pockmarked with ulcers and the road edges begin to be abraded toward the center.

If the blacktop has been omitted from construction, persons using the road and those exercising surveillance do not expect quite as much from the surface and guard the construction more carefully.

Potholes which develop in a laterite wearing surface are usually attended to more readily than potholes which occur in blacktop.

As a planned part of design, an engineer should include education on road surveillance and maintenance as a part of his duties. This is in no way an affront to the ability or perception of the local public works department, but simply sound engineering practice on the part of the designer. He must recognize that he has created a unique construction and knows far better than anyone else the surveillance required in order to observe deficiencies and the means to effect repair when deficiencies occur. If he explains his purposes in this way to the director of public works, his chances of favorable reception are high. If successful, he should establish guidelines for the surveillance program which he thinks would be in the best interest of the road and strive, as diplomatically as possible, to acquaint the director of public works and his road surveillance personnel with the urgency of executing the plan. Most practically, the plan can be effective if it is an integrated part of the surveillance and maintenance program of the entire jurisdiction. Therefore, the design engineer should become thoroughly acquainted with the surveillance and maintenance practices being carried on in the area and, through close cooperation with public works personnel, devise means for insuring initiation of surveillance and maintenance programs which suit the road he designed. The design engineer is cautioned that it takes a keen eye to observe the first symptoms of base-course and subgrade failures in a rain-forest road. By the time the failures are quite obvious, serious damage has occurred to the subgrade which will require extensive repair. An uninitiated engineer should seek means of acquainting himself with the early symptoms of subgrade and base-course failures. The more frequent and obvious symptoms of failures are discussed in this volume. Subtle symptoms of equal significance must not be overlooked. Bleeding of the blacktop surface; appearance of longitudinal, transverse, or diagonal series of spiderweb cracks in the road surface or base course; localized depressions in the wearing surface; unusual tire noise or corduroy surface characteristics; poor surface drainage or the appearance of water through surface cracks from a pumping subgrade; these are all early symptoms of failure and require immediate correction.

Prospective surveillance personnel from the public works department should be carried on field trips by the design engineer. He should point out the various subtle symptoms while at the same time learning from them possible symptoms of which he is unaware and which are unique to the area.

The design engineer should recommend a carefully performed periodic inspection by surveillance personnel. Surveillance inspection on a rain-forest road should be performed, if possible, every week or ten days. The maintenance engineer should plan to perform his own supervisory surveillance inspection at least once a month and the director of public works, overworked though he may be, should conduct a surveillance expedition at least several times during each rainy season.

A procedure of recording surveillance should be devised by the design engineer and presented to the director of public works as being appropriate to this particular construction. A set of the design plans should be used by the surveillance inspector as the inspection work sheet. Locations of impending failures should be referenced from easily recognizable landmarks along the right of way. A narrative record of the reconnaissance should be submitted as a part of the surveillance report, and inspectors should be required to submit a detailed account of locations and observations of symptoms of failure, along with recommendations of the maximum time which can be allowed before repairs should be made without allowing serious damage to the base course and subgrade.

As a part of subsequent client-relations activities, a design engineer should plan to make visits to the site of his construction during several ensuing seasons following opening of the road to make sure that his surveillance suggestions are understood and are being followed, and that adequate maintenance is being performed so as to insure the road will behave properly throughout its intended life. Whatever deficiencies he notes in the surveillance, traffic control, and maintenance programs should be brought to the attention of the director of public works in writing. If financial institutions are involved in financing the project, they too, should be informed.

TRAFFIC CONTROL

The author is familiar with logging operations in West Africa in which the size and length of the logs removed from the forest are governed by the load capacity of the bridges and the turning radius of the curves on the road between the forest and the mill. Except for an occasional bulldozer which the director of public works must trail upcountry to repair damaged subgrade, the logs are the heaviest load the road carries, and they do the greatest amount of damage. Control of this type of traffic is very difficult, but it is less difficult than control of the small privately owned buses or "mammy wagons" which careen down the highway, abrading the shoulders and the edge of the traveled surface.

Traffic control on this particular road which traverses over 70 miles of the most tenuous rain forest in the world, is provided by requiring one-way traffic north one day and south the next day, for six days a week. The seventh day is a day of rest and relaxation for the road, but of hard work for the maintenance crew. This plan follows throughout the dry season—that is, as soon as the road has been put in passable condition following the rainy season. The plan is continued until the beginning of the rains during which the road is closed to all commercial traffic. During the rainy season, the government vehicles which attempt to make the trip leave the roads dotted and scarred with 2-ft ruts totally denuded of wearing surface and sometimes extending into subgrade and base course.

It is not at all necessary that this procedure be followed on a well-designed, ordered, and maintained laterite road in the rain forest, but the well-ordered road requires strict and rigid traffic control. In the event the director of public works does not have the authority to exercise load control on vehicles which use the road, design criteria prove meaningless. Practically, though, load control can usually be accomplished effectively where the owners are willing to include traffic barriers. Careful attention by the designer can result in selection of radii of curves which limit the length of vehicles using the road. Bridge superstructures can be of such configuration as to limit the external dimensions of vehicles. While this seems like an archaic method of exercising traffic control, and to an extent tends to preclude the possibility of upgrading the road at some future date, its effectiveness should not be discounted. If the designer recognizes the value of built-in load-carrying restrictions, he can execute limited curves and include bridges from which restrictions can be easily lifted by increasing the curve radii and removing or modifying the bridge superstructure at some future date should the road subgrade, base course, and wearing surface prove capable of sustaining heavier loads. This foolproof method of protection appears to be the most reasonable to assure the design life of the road.

It must be recognized that the subgrade, once denuded of the base course and wearing surface, is incapable of sustaining design loads. A lack of maintenance and continued vehicular traffic will surely denude the subgrade of these protective elements. It is not sufficient to caution the public works department that total subgrade failure will ensue should traffic be allowed to continue over denuded subgrade. The design engineer is strongly advised to build safeguards into any rain-forest road.

The traffic department of the division of public works is entitled to receive from the designer a sensible traffic-control plan which, if followed, will allow the road to perform adequately while fulfilling its role as a traffic artery. In preparing this plan, the designer should acquaint himself with all types and patterns of vehicular traffic which are currently using the road he is called upon to upgrade. Special attention should be given to other roads in the region of higher order so that he may visualize the type of traffic which his new road might be expected to receive immediately. Furthermore he should logically project load and vehicular characteristics which may be expected within the design life of the road and his plan of traffic control should take these items into consideration. He should plan a realistic traffic standard to effect the proper use of the road. Further suggestions as to specifics seem superfluous, for each set of circumstances and each economic pattern will be different. In the event the designer questions the methods of outlining such a plan he should avail himself of the services of a consulting transportation or traffic engineer.

TREATMENT

The well-designed, constructed, and disciplined road or airfield may be expected to resist the abrasion of traffic and the effect of weather during its design life. It must be anticipated, however, that the surface will become abraded, the edges will be eroded by rains and damaged by wheels, the shoulders will be washed away or totally destroyed by wheel rutting, and drainage will become clogged or washed away through the accumulation of debris and the ensuing redirection of water. Treatment must be anticipated in the maintenance program and carried out before damage occurs which will necessitate repairs.

Special attention is directed to drainage structures as well as to natural and artificial drainage courses. As these courses are part of the initial design, their continued function is essential if water is to be controlled. Loose debris, fallen timber, branches, boulders, and other *large* objects are serious enemies of continued effective drainage since it is these that are often lodged in drainage courses forming the nucleus of a dam which may eventually direct water to a place where it will severely damage the construction.

Surveillance teams must be schooled to recognize the danger signals of clogged or impeded drainage in order to relieve the stoppage. The inspection of drainage structures and courses must be continuous during the rainy season and where stoppages are noted the offending material must be removed. The articles which cause stoppage are mostly large and not amenable to removal by a small maintenance crew armed with hand tools. Often bulldozers can be used effectively to remove tree trunks and large boulders, but unless the maintenance crew uses extreme care more damage than benefit may result from their use. Therefore, the use of such equipment is indicated only when it is clear that other methods cannot be employed in a timely and safe manner.

More appropriate is the use of mechanized chain saws to reduce tree trunks to manageable size for removal. Likewise, a few well-placed sticks of dynamite can break boulders to one-man-sized rocks.

Grassed shoulders and slopes must be replanted, strewn with straw, oil treated, or matted when and where they are most likely to erode.

Usually there is an antierosion technique used locally that includes readily available materials. Check barriers of forest timber or boulder riprap may be used to prevent slope erosion.

Where construction traverses terrain containing volcanic ash and lateritic soils, slopes are frequently designed vertical as these materials stand best on such cuts. It has been observed that vegetation frequently grows above these slopes and hangs down over the cut face. This vegetation both overloads the unprotected face and provides root paths for surface water to percolate into the soil causing face sloughing. This may be prevented by cutting the surface vegetation away from the top of slope; in no case should vegetation be allowed to hang down a vertical cut face even though it may appear to be affording protection. This is believed to be the most important requirement in brush cutting and must be carried on continuously as contrasted with the yearly brush cutting on cleared right of way.

The treatment of wearing surfaces should be planned and executed on schedule regardless of the *appearance* of the surface to be treated. For compacted wearing surfaces of laterite the surface may be rejuvenated by a shallow scarifying (1–2 in.) upon which is placed a well-crushed blanket of laterite. The scarified material and the newly placed blanket should be thoroughly compacted with the smooth-wheel roller, taking care to assure that the design crown and curve superelevation are maintained.

Treated laterite surfaces should be retreated every two years at a minimum. In the event the surface has been abraded it should be returned to original grade by the introduction of additional material. Single shots of cutback asphalt may be employed repeatedly if it is necessary that the road be kept open during treatments. Such a mixture cures readily and quickly during dry weather but is of little value if employed during rainy seasons.

REPAIRS

In the great majority of cases surface failures are the result of a failure of the subgrade. The major cause of subgrade failure is a severe loss of soil strength due to an increase in moisture content beyond tolerable limits. The major cause of an increase in soil moisture is the result of improper subgrade drainage and the introduction of soil moisture from below the zone of failure.

An understanding of the behavior of surface and groundwater can serve

Fig. 23. Failed subgrade of West African road.

as a meaningful beginning in preventing or repairing construction failures. Surface drainage is important and must be maintained through crowning, ditching, and culverting. Impermeable barriers such as the wearing surface must be kept in integrity so as to prevent downward percolation of water. But water which drains through soil will seldom do the damage that will water which *remains* in soil. Where failure occurs, a waterlogged soil must be the first suspect.

Immediate repairs should be undertaken to prevent deterioration or destruction of the construction. Concurrently, the real cause of the failure should be researched.

Subsurface water percolating upward through a subgrade must be identified and its source discovered. Often a single failure along a construction section is but prelude to many failures within the general area. Figure 23 shows such a failed area. Small-diameter auger borings will usually reveal the extent of the saturated subgrade soil at the location of a failure. Slope-face seepage may be a further clue to the extent of the saturated zone. Further borings peripheral to the initial failure will serve to define the limits of the saturated soil.

Once defined, the source of the subsurface water may be traced usually to a geologic or topographic feature which is directing water into the earth mass through hydrostatic pressure. Subsurface drainage through interception ditches or French-type underdrains is the simplest method of diverting subsurface water. If the offending aquifer is found to be fractured supporting rock, drainage can sometimes be affected and pressure released by locating and opening a drainage path at an elevation *below* (downslope from) the

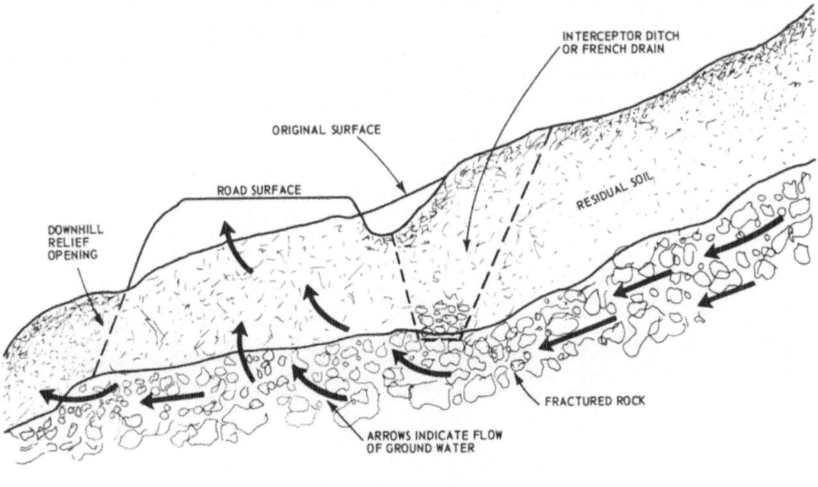

Fig. 24. Principles of subgrade drainage.

saturated soil mass. If water percolating through fractured rock is allowed to issue forth at an exposed face in the fracture zone, it will usually seek the path of least resistance which is the freely flowing rock. Of course, if upslope soil is found to be the source of groundwater, it is usually best to install drainage upslope of the soil mass which must be protected. In Fig. 24 these principles are illustrated.

To correct construction failures in a wearing surface, the entire mass of damaged subgrade soil must be defined, removed and replaced with soil of appropriate strength. Often this will require the removal of a considerable portion of the subgrade. This is best accomplished by removing the failed wearing surface and the immediate damaged base. Probing with a slender steel rod will reveal soil which is soft and moist. This damaged soil should be removed at least to the limits of practicality, which, at a minimum, is laterally and vertically to the depth of influence of the applied design load.

If subgrade damage is the result of upward water infiltration, provisions must be made to relieve a future moisture condition by underdrainage. If, however, surface water is the cause, replacement with an unyielding subgrade material covered with an impermeable surfacing will suffice. Replacement of damaged subgrade soil is frequently accomplished during rainy weather when borrow soil may be at a high moisture content. In order to forestall difficulty in lowering soil moisture, it is often desirable to use a granular material or other similar soil in which excess moisture does not inhibit achieving a high density by compaction. Sand, crushed laterite, or granular volcanic materials are suitable for this purpose. The materials should be placed in the prepared cavity in 6-in. to 1-ft lifts and hand tamped to a dense state. Where grade is reached, an impermeable base-course blanket of laterite should be constructed with particular attention to the juncture of the replacement with the adjacent base. The new wearing surface should extend at least 1 ft beyond the perimeter of the replaced soil mass and should be well fixed and, if possible, joined to the surrounding surface by traversing the newly surfaced area by multiple passes with a smooth-wheel roller.

Appendix

EXCERPT FROM

SOIL MECHANICS TEXTBOOK U. S. ARMY ENGINEER SCHOOL FT. BELVOIR, VIRGINIA

CHAPTER XI: LATERITES AND LATERITIC SOILS OF VIETNAM

Introduction

The purpose of this reference is to define and discuss laterites and lateritic soils and to present the characteristics and considerations necessary for their successful use as a construction material.

Laterite is a porous, indurated, concretionary material which is usually red to reddish brown in color. The name laterite is derived from the Latin word "later" which means "brick earth." This material was first observed and described by Buchanan in 1807 during his travel through India where laterite was being quarried from the subsoil stratum, trimmed, and used as building blocks for local buildings. There are three types of laterite commonly encountered in a tropical climate. The names of these types describe their physical appearance. Wormhole laterite (vermicular) is a massive concretionary formation with an iron-rich matrix and a slaggy or wormhole-like appearance. Pellet laterite (oolitic) consists of fine soil grains which are cemented by iron oxide into pellet-shaped particles. These pellets may be loosely consolidated or unconsolidated. The third type of laterite is a "soft-doughy" material which hardens irreversibly upon exposure to alternate wetting and drying. This type of laterite is found extensively in Africa but does not exist in significant quantities in Southeast Asia. Since this volume considers primarily laterite soils of Southeast Asia, specifically South Vietnam, the characteristics of "soft-doughy" laterite will not be discussed. Wormhole and pellet laterite will also become irreversibly harder and more stable upon exposure to alternate wetting and drying. This unique property supplements the physical properties to make these two types of laterite desirable construction materials and an important group in the tropical soils family.

Laterite is found in the soil profiles of sparsely vegetated and rolling to roughly dissected terrain. It is formed through the action of minerals from weathered igneous rocks which are carried into the subsoil stratum by a fluctuating groundwater table.

Laterite is often confused with lateritic soils simply because the physical appearances and chemical properties are so similar. Lateritic soils vary in type from poorly graded sands (SP) to highly plastic clays (CH) and vary in color from a red to a reddish brown. These soils characteristically exhibit some secondary iron cementation between mineral grains; however, there is a wide variation in the degree of cementation. Although lateritic soils also harden upon drying, they soften readily upon wetting. The lateritic soils are finer grained materials than laterite and behave accordingly. An important physical difference between a laterite and a lateritic soil, is that a laterite has a gravel component but a lateritic soil does not.

It is often very difficult to determine whether a sample is a laterite, a lateritic soil, or a tropical red gravel. However, the differences in behavior of these kinds of soil material are significant and erroneous classification could lead to serious construction failures and/or hazardous consequences. For example, decomposed red granites are often mistaken for laterite and used as a base course for a road. Unfortunately, the first rainfall will turn such a roadway to mud and render it useless if traffic is allowed on it before it dries sufficiently.

Environmental Characteristics

Climate

A prerequisite for the formation of a laterite and/or a lateritic soil is a climate which is both tropical and monsoonal. The characteristics of this climate are: long alternating wet and dry seasons with short transitional periods, a constant annual temperature of about 85°F, and a high amount of rainfall during the wet season. These characteristics result in high cyclic fluctuation of the groundwater table and accelerate chemical weathering.

Hydrology

In order for lateritization to occur, the surface and subsurface water must be iron rich and chemically active. The soil profile should be porous enough to allow the groundwater table to fluctuate within the subsoil. The groundwater table characteristically exhibits a distinct high level and low level with the wet and dry seasons respectively. The high level of the water table must be located a minimum depth of 3 ft below the ground surface in order for oxidation and lateritization to occur. The soil is supplied with iron from the rejuvenating surface water, seasonal fluctuation of the groundwater table, seepage from adjacent higher terrain, and percolation. The fluctuating of the groundwater table aids in the lateritization process. Typically, the level of the groundwater table within mature lateritic soil profiles is either directly below the laterite or at a shallow depth in the white clayey-silt substrata. Sometimes a perched water table is encountered above a laterite formation.

Landforms

Lateritization commonly occurs in the soil profiles of a plain, plateau, and/or terrace. Laterite is found only in mature terrain which is characterized by moderate to heavy dissection. Typically, the terrain is rolling to roughly dissected with numerous hills and hillocks. Breaks in slope, scarps, interfluves, and flattened hilltops within this terrain exhibit outcrops, and are good potential locations of laterite. Laterite is most likely to be located on the older and flatter slopes of the hills and dissected features.

Lateritic soils are typically found either above laterite deposits or alone on younger, moderately dissected terrain. This terrain is usually undulating to rolling with some hillocks. The red and brown clays are found on young, low, flat to undulating terrain. Low flat areas such as the Mekong Delta generally are continuously too wet for the lateritization process to take place.

The upper limit for a landform as a probable source of laterite is a hill which has been uplifted and dissected from a plain, and not eroded from a mountain. Typically, the hills and especially the mountains preclude the formation of a laterite because of their unsatisfactory climatic and hydrological characteristics. Therefore, the foothills and mountains of the I and II Corps sectors in Vietnam cannot be considered as sources of laterite except for small local outcrops. Soils in this area which are commonly mistaken for laterite are either decomposed granites or dry dark-brown silts.

Geology

The type of laterite formed depends upon the type of rock being weathered and the amount of iron available. For example, laterite derived from a basaltic rock is usually thick, hard, and dense. Laterite is formed from iron-rich basic rocks such as basalt, granite, and gneiss granites. It is not derived from metamorphic rocks. Laterite is formed as a residual soil (in place). Although there are extensive areas of pellet laterite in Thailand above sandstone, laterite formations above sandstone are thought to be more of an "accident of location" than a natural formation. The hydrolytic weathering of basic rock minerals such as the ferromagnesium silicates (biotite, hornblende, phroxene) and plagioclase and orthoclase feldspars yields the necessary iron and aluminum for lateritization. The weathered rocks below the layer undergoing lateritization and from adjacent higher areas are two principal sources of these minerals.

Soils

The surface soil above a laterite formation is usually hard, uneven, highly leached, red to reddish brown (even black), and has black, glossy iron pellets scattered locally. This soil is well to moderately well drained. The laterite formation below is horizontally discontinuous in extent or dendritic and varies in thickness from 6 in. to 30 ft. Wormhole laterite usually forms a much thicker formation than pellet laterite. Laterite may be found as an outcrop or in deposits as deep as 30 ft below the surface and is formed from soils of the Tertiary Period; the lateritic soils are younger.

Vegetation

Vegetation above a laterite or lateritic soil usually is light to sparse. Cassava, rubber, brushwoods with some scattered trees, or bamboo within forrested areas are typical. When laterite is near the surface, vegetation normally is limited to low grass. Deeper laterite or lateritic soil deposits permit a denser vegetative cover.

Lateritization

Wet Season

During the wet season rainfall greatly exceeds evaporation. The heavy rainfall becomes alkaline (pH of about 8) and it readily hydrolyzes the silicates in the rock. Some of the rock minerals hydrolyzed are biotite, hornblende, pyroxene, and the feldspars. Most of the iron, aluminum, magnesium, sodium, and calcium ions are leached from the rock minerals. The divalent metal ions in solution in the groundwater flocculate the silica and preserve a high silica to alumina content within the soil-water system. Hydrated alumina is insoluble at this pH and precipitates as hydrated gibbsite. If the silica is immobilized, it will precipitate from the solution; a kaolin clay will be formed when a 1:1 ratio of silica to gibbsite is maintained. However, if this ratio is greater than 1:1, the insoluble aluminum silicate may be deposited separately as hydrated alumina and silica.

Highly soluble and unstable ferrous (Fe^{++}) ions are carried from the weathered rock to the subsoil by groundwater which becomes increasingly more basic as time progresses. These iron ions are in the form of ferrous (Fe^{++}) bicarbonate or a complex ferrous hydroxide. However, the iron will be oxidized to a ferric (Fe^{+++}) oxide in areas where no humic acids and other organic products can significantly change the pH of the environment. If the environment or groundwater remains acidic, iron oxidation will be prevented. Once the iron is oxidized, it precipitates from solution as a ferric oxide, neutralizes the cation-exchange capacity of the soil particles by saturation, and eventually forms a "red mottle" of hematite. As precipitation continues, the red mottles soon are increased into cores and finally into a laterite deposit.

When exposed to alternate wetting and drying, the oxidized and precipitated iron segregates and finally crystallizes.

Transitional Period

During this period, the moderate rainfall is sufficient to remove most of the iron ions, but not heavy enough to flush away the hydrolyzed silicates, the hydrogen ions from plant acids, and organic clays. The pH of the groundwater is acidic (about 4 to 5), which is sufficiently low to make both the aluminas and silicas relatively insoluble. However, the combination of the hydrogen or iron ions with the alumina and silica can result in the formation of a kaolinite clay if the ratio of silica to alumina is 1:1. The acid environment induces kaolinization by readily substituting the hydrogen ion for the iron ion by cation-exchange means in the clay mineral. The displaced iron ions will migrate to the alkaline environment of oxidized and precipitated iron mottles. This iron deposit is both chemically and electrically attractive. Here the iron ion will then be oxidized and precipitated from solution. Upon exposure to alternate wetting and drying, the precipitated iron will eventually crystallize. Gibbsite deposited during the wet season will be resilicated to kaolinite during this transitional period.

Dry Season

During the dry season evaporation exceeds rainfall. However, some hydrolytic weathering does occur. There is relatively little leaching of sodium, calcium, potassium, magnesium, aluminum, and iron ions from the rock since the rainwater usually is evaporated before it can carry anything away in solution, and the water table shows little fluctuation. The alumina and silica obtained from hydrolytic weathering of the rock minerals are immobilized by flocculation and insufficient rainfall. The groundwater usually is alkaline with a pH greater than 7.0. The solubility of both the silica and alumina derived from the parent rock is relatively low at this pH. They will combine to form kaolinite clay if the silica to alumina ratio of the precipitate is 1:1. The metal ions which control the degree of flocculation of the silica in solution constantly are being oxidized or are combining to form clay minerals such as kaolinite. A montmorillonitic clay or illitic clay will form if the groundwater table becomes stable during this period, if calcium and potassium ions are readily available, and if there is an excess precipitation of silica. This season readily induces kaolinization and allows for some iron oxidation.

Development

The horizontal seepage and periodic vertical fluctuation of the groundwater table causes the clay to migrate to the lower stratum and the iron mottles to eventually develop into laterite deposits. The alternate wetting and drying of the soil eventually will cause crystallization of the precipitated iron. During the crystallization process the iron ion complexes are segregated and reduced to their lowest energy state. As the oxidation, precipitation, and crystallization processes continue, the red mottled soil is transformed into a lateritic soil and eventually into a laterite. This process is called lateritization. Wormhole laterite usually is formed in coarse-grained alluvial soils and pellet laterite is formed in fine-grained soils. The ultimate product of lateritization is either hematite or bauxite.

Occurrence in RVN

Although any red soil found in RVN usually is considered to be a laterite, only about 30% of the country satisfies the conditions required for the formation of a laterite. Massive formations of wormhole laterite are found in the southern portion of the Mekong Terrace which is an east–west band just north of Saigon. Pellet laterite is found in the northern portion of the Mekong Terrace and on the Ban Me Thuot and Pleiku Plateaus. The northeast coastal lowlands which extend from Qui Nhon north to Hue are good probable sources of either type of laterite.

Laterite should not be found in the soil profiles of the following areas: The Mekong Delta (it is too low in elevation and too wet); the foothills and mountains of I and II Corps Tactical Zones (they are too cold and/or dry); and the coastal lowlands of Phan Ranh and Phan Thiet (they are shielded from the monsoonal winds by the Annamite Mountains). However, within these areas laterite may occur locally in limited quantities. The red soils commonly encountered within the hilly and mountainous areas, although often considered to be laterite, usually are decomposed granites* or dry dark-brown silt.

Typical Laterite Profiles

Pellet Laterite

1 ft	(1)	Silty clay or sand.	
1 ft	(2)	Dark red to yellow clay (sometimes present).	
5 ft	(3)	Lateritic soil.	
3 ft	(4)	Pellet laterite.	GWT
Deep	(5)	White clayey silt.	

(1) Hard, silty clay or silty sand; highly leached; and containing some organic matter with black, glossy iron pellets scattered locally.

* For example, in the "Seven Mountains" in Chau Duc Province and the "Three Ladies" in Kien Giang Province (northwestern portion of the Mekong Delta).

(2) Dark red to yellow mottled clay (this layer may or may not be present).
(3) Lateritic soil may classify as soil-type SP or CH using the Unified Soil Classification System. It is usually red to reddish brown in color.
(4) Pellet laterite may be loosely consolidated or unconsolidated pellets in a soil matrix. It is usually red to reddish brown in color.
(5) White clayey silt with red mottles and highly compressible (MH).

Wormhole Laterite

2 ft (1) Silty clay or sand.
15 ft (2) Wormhole laterite. GWT
Deep (3) White clayey silt.

(1) Red to brown or tan silty clay with some fine sand (CL).
(2) Wormhole laterite (secondary rock) red to reddish brown in color.
(3) White clayey silt with red mottles and highly compressible (MH).

Keys to Photo Interpretation

The following keys (or clues) to the presence of laterite should be considered in the order presented to complement and substantiate the airphoto interpreter's decision regarding the possible location of laterites.

1. Landform—rolling to roughly dissected plain, terrace, or plateau with hills and hillocks locally.
2. Age—mature terrain characterized by numerous dissected features.
3. Vegetation—usually sparse and comprises cassava, rubber plantations, shifting cultivation, brushwoods, or bamboo.
4. Drainage—well to moderately well drained. Golf courses and cemeteries are good probable locations of laterite.
5. Soil surface—hard, bumpy, red to reddish-brown soil which appears dark and coarse textured on an aerial photo.
6. Relief—scarps, interfluves, and flattened hilltops may indicate laterite outcrops.
7. Cultural features—the consideration of most cultural features requires coordination between ground reconnaissance and aerial photo interpretation. Road cuts, earthen-block buildings, and hand-dug wells will give an indication of locally present laterites.

Physical Properties

Wormhole Laterite—Vermicular

Wormhole laterite in South Vietnam has a red to reddish-brown color and a slaggy or wormhole-like appearance. It is a massive concretionary formation with iron-rich cores sometimes interconnected by iron-rich lenses. The specific gravity ranges from 2.76 to 3.50. Due to the numerous interconnected holes, it usually is well drained with a permeability (internal drainage) much like that of a clean sand or gravel. This secondary rock upon exposure and drying will harden irreversibly to a MOH's scale of hardness of about 3. When excavated by conventional methods, a typical percentage of soil components would be above 40% gravel, 30% sand, and usually less than 30% fines. Typically the Atterberg Limits for the minus #40 sieve material are: a liquid limit from 30 to 50%, a plastic limit from 10 to 20%, and a plasticity index of from 20 to 30%. The Unified Soil Classification System classified this material as a clayey or silty gravel (GC or GM). Since this laterite commonly occurs in the coarse-grained alluvial soils of the Mekong Terrace, it is often referred to as "Groundwater Laterite" or "Bien Hoa Stone."

Pellet Laterite—Oolitic

Pellet laterite is red to reddish brown in color and is commonly found above or near bedrock (e.g., the Ban Me Thuot Plateau). This type of laterite consists of fine-grained soils which are highly iron cemented into pellet-sized particles. When these pellets are broken in half, a very fine-grained soil matrix may be observed. These pellets can be found loosely cemented with other fine-grained materials to form a conglomerate rock, or they can be found as uncemented gravelly soil with a high percentage of fines. The wormhole and pellet laterites are highly resistant to chemical and mechanical weathering. When excavated by conventional methods, pellet laterite exhibits the following percentage of soil components: approximately 40% gravel, 10% sand, and usually less than 50% fines. The Atterberg Limits for the minus #40 sieve material are: a liquid limit from 30 to 50%, a plastic limit from 25 to 35%, and a plasticity index from 5 to 25%. The pellet laterite has a higher plastic limit and a lower plasticity index than the wormhole laterite. This material is classified as a GC or GM based on the Unified Soil Classification criteria and has a specific gravity which varies from 2.76 to 3.5. The maximum particle size varies from a high of 10 in. to a low of 1/4 in. When the material is exposed to alternate wetting and drying, it will become irreversibly harder. Be careful not to confuse this pellet gravel with the highly weathered and unstable tropical red gravel. Tropical red gravels are a hazardous construction material and should be avoided at all times.

Lateritic Soils

Lateritic soils are red to reddish brown in color and vary in type from a poorly graded sand to a highly plastic clay. These soils have some secondary iron cementing between particles and a specific gravity of from 2.73 to 3.12. When this material is undisturbed, it is fairly well drained; however, after being disturbed it is relatively impermeable and plastic. Typically these soils excavate as a fine-grained material with a percentage of soil components of about 40 to 50% sand, 30 to 40% silt, and 20 to 30% clay. The Atterberg Limits for the minus #40 sieve material are: a liquid limit of 40 to 70%, a plastic limit from 25 to 50%, and a plasticity index from 15 to 20%. Since the liquid and plastic limits are equally high for a laterite, the plasticity index is about the same for a laterite and lateritic soil. The significant difference between a laterite and a lateritic soil is the presence of gravel components in a laterite and their possible absence in a lateritic soil. An important physical consideration of a lateritic soil is that it will not harden irreversibly upon drying.

Chemical Properties

Laterite

One of the most important distinguishing chemical characteristics of a laterite is the silicon dioxide to ferric oxide ratio (SiO_2/Fe_2O_3) which must be less than 1.33. A laterite will have a lower base-exchange capacity and higher pH than a lateritic soil. Although kaolinite is the predominant clay mineral, unavoidably there will be some illite and montmorillonite present. The iron content of the bulk sample usually is 5 to 10%, while the iron content of the cores in wormhole laterite and pellet laterite may be as high as 20 to 40%. The progressive hardening of laterite when exposed to alternate wetting and drying will occur over a relatively long period of time because the outside crust hardens first, retarding the hardening of the inner core.

Lateritic Soils

The lateritic soils have a silicon dioxide to ferric oxide ratio (SiO_2/Fe_2O_3) greater than 1.33 but less than 2.0. The base-exchange capacity is greater and the pH is lower for a lateritic soil than for a laterite. The clay minerals usually are kaolinite, illite, montmorillonite, or any combination of the three. The iron content (0–2%) of a bulk sample of a lateritic soil usually is less than that of a laterite. When this solid dries, the iron forms a weak, reversible secondary bonding between particles. The combined effect of the secondary iron bonding and the clay hardening upon drying gives this soil a high dry strength. Since a lateritic soil may exhibit physical appearances and chemical properties similar to a laterite, a clear distinction between the two may sometimes be relatively difficult. In this case it is not important whether the material is a laterite or lateritic soil, but how suitable it will be for the particular construction purpose.

Red and Brown Clays

These clays have a silica to ferric oxide ratio (SiO_2/Fe_2O_3) greater than 2.0. Their pH is lower than that of a laterite and lateritic soil. However, the base-exchange capacity for the red and brown clays is higher than that of a laterite.

Testing

Laboratory Testing

The military engineer in his evaluation of laterites and lateritic soils must determine their suitability as construction materials by determining their physical and chemical characteristics.

Standard laboratory tests properly executed will indicate (1) the suitability of material for a specific construction purpose and (2) the degree of lateritization.

The physical properties should be determined first. The sieve analysis will indicate percentages of the soil components and their gradation. It is important to thoroughly wash the material being tested through all sieves including the #200. A dry sieve analysis will not indicate the correct total amount of fines. The material should be soaked for 48 hours prior to performing the Atterberg tests.

The coarser grained material usually has a higher degree of lateritization and is more desirable. Comparison of the typical results of these tests for laterites and lateritic soils is an initial step in identification and differentiation. Soils with low liquid limits, plastic limits, and swell characteristics generally are more desirable for use in construction. Refer to *Soils Engineering, Volume I* for test details. One of the best laboratory tests for determining the suitability for construction and degree of lateritization of a sample is the Modified AASHO Compaction Test. Although the compaction curves obtained for a laterite and lateritic soil are similar in shape, they are significantly displaced one from another. Both curves are asymmetrical about the optimum moisture content; they are much more steeply sloped on the wet side. A laterite typically has a maximum dry density of from 130 to 150 lb/ft^3 and an optimum moisture content of from 8 to 12%; a lateritic soil has a maximum dry density between 110 to 120 lb/ft^3 and an optimum moisture content of 12 to 16%. Therefore, the compaction curve for a lateritic soil is lower and shifted to the right of the laterite compaction curve.

The CBR test on laboratory samples will indicate the degree of lateritization of a soil. The CBR for a laterite is highest just dry of the optimum moisture content and increases with an increase in the percent of gravel and decrease in the percent of fines. A laterite sample soaked for 3 to 4 days at optimum molding moisture content will have a greater strength than a sample soaked at a moisture content less than optimum. Generally, the greater the CBR value, the higher the degree of lateritization.

After an initial air-dry moisture content is determined, the dry sample should be weighed and placed in an oven at 500°F for at least 48 h. The water lost in this test would be water of crystallization. The greater the loss of water, the greater the degree of crystallization or lateritization. The slaking test will also aid in determining the degree of lateritization and the suitability for construction. If a ball of minus #40-sieve material submerged in water disintegrates immediately, it is undesirable for construction and has relatively little lateritization. If the sample disintegrates slowly during a 24-hour submergence, it usually is suitable as a construction material. If time is available, the process of drying and soaking should be repeated to see if the sample changes significantly from cycle to cycle. A laterite will exhibit an increase in hardness with alternate wetting and drying, while a lateritic soil will progressively degrade. An abrasion test will indicate the degree of lateritization. The higher the degree of lateritization the less the sample will abrade. Since a Los Angeles abrasion device usually will not be available, some expedient device such as a closed cylinder with steel balls should be used and the procedure standardized. Then the results for a laterite can be compared with the sample in question. If the sample has less than 1% montmorillonite and less than 20% of other clay minerals there will be little shrink or swell.

Field Tests

In the field as in the laboratory the physical characteristics must be determined as extensively as possible. Field identification tests such as sieve analysis (on whole soil) and dry strength, ribbon, wet shaking (dilatancy), bite, grit, and shine (all on minus #40 sieve fraction) should enable the soil analyst to classify the soil and get an idea of the relative degree of lateritization. The hasty sedimentation test will serve to indicate the degree of secondary iron cementation of the fine-grained soil particles. From these test results and previous experience, a judgment can be made on how suitable this material is for construction. These field identification tests should be supplemented with laboratory tests whenever possible even though some crude attempt to duplicate a laboratory test must be made in the field.

Construction Procedures

Borrow Site

In developing the layout of a borrow site, area utilization and drainage (especially during the rainy season) are key factors to consider. If scrapers are to be used, the borrow pit should be excavated from an uphill position down, and the furrows made by the scrapers should be continuous and provide for drainage away from the pit. However, if power shovels are to be used, the pit should be excavated from a downhill position up. This technique will permit natural drainage and prevent local ponding. All stripped soil should be placed in an area without deposits and which is not expected to be used later. While the laterite is being excavated, be sure that no underlying white clayey silt is mixed in through overcutting. If silt is mixed in during excavation, it must be washed out prior to using the laterite, otherwise serious local failures may be expected. The lateritic soil usually can be excavated with a scraper or a scraper pushed by a bulldozer. However, the laterite must be excavated with a bulldozer with a 12- to 18-in. ripper tooth. Blasting to excavate laterite is relatively impractical since the laterite has a high natural porosity.

Field Compaction

Laterite. During excavation, transportation, and compaction, an effort should be made to prevent unnecessary structural degradation of the laterite; therefore, compaction should be light and shaping kept to a minimum to avoid high shear stresses. For wormhole laterite, an 8- to 10-ton, vibratory, steel-wheeled roller gives the best result. For pellet laterite, the 5- to 8-ton steel-wheeled or pneumatic-tire rollers are the most effective.

Lateritic Soils. In Thailand, the Hyster Grid Roller (used by the Navy) was used effectively in compacting a lateritic soil. The roller is similar to a sheepsfoot roller except that the feet are flatter and have a larger surface area. It has the capability of compacting thick lifts; for example, a 12 in. loose lift can be compacted to a 6 in. compacted thickness. However, the loose-lift thickness is usually limited to 6 in. and is compacted to a thickness of 3–4 in. Contrary to popular opinion, the sheepsfoot roller can be used effectively if the weight of the roller is reduced (usually by only half filling the drum) and the roller is pulled slowly to avoid high shear stresses. The 50-ton roller can be used, but the load and tire pressure are critical; they must be adjusted to approximately 25 tons and 90 to 110 psi respectively. Lateritic soil compacted on the wet side of OMC often will give a spongy section instead of a suitable compacted layer. A good rule of thumb for the field is to apply water at 2% less than the lab optimum moisture content.

Field Density. Test samples should be taken at randomly selected points across and throughout the length of roads, landing strips, and taxiways to insure that the desired conditions are being achieved. The density of these samples should be determined by the sand-cone method. This method is the quickest and most widely used. Samples should be taken from dry, wet, and average roadway sections. The moisture content of the sample can be determined readily by the Speedy Moisture Content Device. If this device is not available, expedient methods such as burning the soil in alcohol or just cooking it in a mess kit are acceptable. The field compaction curve obtained usually is similar to the lab curve, but displaced to the left and maybe slightly higher. Frequent field density tests should be performed to insure that the percent of compaction required is being achieved; these field density tests should be compared to the lab tests to insure that no new material has been encountered. A simple field test to approximate the optimum water content for compacting a lateritic soil in the field is to squeeze a sample of soil in your hand; if a slight impression of the fingers is retained in the soil, it is near optimum moisture content. An expedient check on the degree of compaction in the field is to drive a nail into the compacted layer. If the nail gives a hard, metallic, resistant ping to the driving effort, then the compaction is satisfactory. However, if the nail drives into the compacted layer rapidly, the compactive effort is insufficient. The nail test assumes that the soil was compacted at or near design moisture content. If the soil were considerably dryer than OMC, the surface layer would become "pinging hard" but the underlying part of the lift would be insufficiently compacted.

Construction Precautions. A roadway should be elevated a minimum of 4–5 ft, and a runway should be elevated about 7–8 ft above the groundwater table. Whenever practical, the roadway should be constructed above the annual flood level, or adequate drainage facilities constructed. During the placing of fill material, keep the roadway well crowned. During the rainy season this will prevent serious local ponding and structural softening. In all cases, adequate drainage ditches, culverts, and interceptor ditches must be provided to handle the intense runoff. This is a key factor to insuring the long-term use of a roadway or airfield. During the dry season, the fill may be depressed at the center of the roadway to improve compaction operations; however, it must be recrowned at night in case of rain. The compacted layer should be scarified before placing the next layer to permit good interlayer bonding. A crown of 4% is necessary for an unsurfaced roadway; a 2% crown is satisfactory for surfaced roadways or landing strips. The layers should be sealed immediately after compaction. The wobblewheeled roller and pneumatic-tire rollers such as the 50-ton roller are effective. A graded filter blanket 4–8 in. thick between the subbase and subgrade is effective in interrupting capillary action and draining the subbase. Thicker blankets should be used if time permits. Roadway and landing strip side slopes should be kept clean and shaped. Slopes of 3:1 are permissible, but slopes of 8:1 are required often for stability; slopes up to a maximum of 2:1 will support vegetation.

Dust is a major problem on lateritic roads and airfields. If a source of salt water is near it can be used as an excellent expedient dust palliative and soil stabilizer. Asphalt is another very effective dust pallative. The application of 0.5 gal/yd^2 of medium- or slow-cure asphalt cutback (MC-30, 70, or SC-70) upon a premoistened roadway is most effective, giving a penetration of up to 3/4 in. Dry laterite should be spread over this asphalt and a second application of asphalt cutback applied. This procedure yields a satisfactory and durable temporary surface course. Periodic reapplication of alternate layers of asphalt and laterite will insure a fairly long functional life.

Stabilization

General

Laterite and lateritic soils are stabilized to prevent dusting and to provide waterproofing. In addition, lateritic soils are stabilized to increase their strength and stability. Lime, cement, and asphalt are three common stabilizers used. If these stabilizers are used, laboratory testing should be performed to insure that the design criteria can be achieved and the overall stabilization effort is well worth the time and money. Since laterite excavates as a coarse-grained material with a high porosity and a large exposed surface area, stabilization is primarily concerned with the fines component. A lateritic soil varies from an SP to CH and, therefore, the stabilization problems are quite varied and often very troublesome. In general, the higher the degree of lateritization, the higher the response to stabilization. This is a result of a more stable iron-cemented secondary structure and a decrease in the affinity between a lateritic soil and water. Care must be taken not to add too little or too much stabilizer to a soil; the respective results will be either no effect or reduced strength and instability. A minimum of 2% for lime, 3% for cement, and 5% for asphalt must be added to a soil before desired results can be realized. Adding too much asphalt or lime must be prevented, especially when stabilizing a silt.

The volume of stabilizers added should be absolute and determined from their known specific gravities and weights. These volumes should be checked with the total volume measured to make sure no miscalculations were made. If the organic matter present in a soil is greater than 5%, or less than 5% but containing certain types of organic matter, there may be serious stabilizing problems. Before attempting to stabilize a soil, a check should be made to insure that the soil does not exhibit significant swell and shrink characteristics.

Soil Cement and Lime

To determine the optimum moisture content and percent of stabilizer to be added to a soil, a Modified AASHO test must be performed on the natural soil. Once the optimum moisture content of the natural soil has been determined, various samples should be compacted at different lime or cement contents, at water contents which bracket the optimum moisture content by ±4%. For a particular density required by design, a satisfactory combination of water content and percent stabilizer should be chosen which is compatible with economy of material and ease of placement. To estimate the water content required in the field for a soil cement, add to the optimum moisture content of the natural soil an amount of water relative to concrete using a water to cement ratio of 5 gal/ft^3 of 1/2 the weight of cement in water. The soil-cement compaction curve is a flattened parabola and the greatest water content compatible with placement should be added. A good estimate of the water content to be added to a lime stabilized soil is 2% more than the optimum moisture content. Generally, stabilization of laterite with cement is best done with about 4 to 8% by dry weight of cement. Lime stabilization of a laterite is not recommended. Stabilization of lateritic soils is best accomplished with 4 to 10% of quick lime; however, the same proportion of soil cement has been used with satisfactory results. However, 8 to 14% of a hydrated lime is required for satisfactory stabilization. The curing period for soil cement is 1 to 3 days, while a minimum of 7 days is required for a lime. Cement particles will penetrate more rapidly than lime. It cannot be overemphasized that sufficient water for cement curing must be added. The formulae for the amount of lime or cement to be added to a soil for a given percent of stabilizer are shown in the previous chapter in this volume.

The cement or lime should be placed and mixed with the soil dry. Then as much water as would allow the equipment to effectively mix the components and work the area should be added. After mixing, the rest of the water should be added, the components mixed again, and then finally compacted. The soils should be compacted as soon as possible after mixing. The longer the delay between mixing and compacting the less the final dry density will be. For a soil cement, this delay should not be longer than 1 hour, while for a lime 2–3 hours are permissible. Lime is often applied to a soil as a slurry since it prevents the burning of eyes and skin irritation. The lime should not be overexposed to the air by excessive handling. For a quick lime, the soil layer should be moistened after compaction to facilitate curing.

Unconfined compression tests should be performed on samples 2 in. in diameter and 2 in. high to insure that the compressive strength required is achieved. If a loading frame is not available, a good expedient test could be performed by placing a sample on a bathroom scale and pushing down on the top of the sample. The load applied can be read directly from the bathroom scale.

Asphalt

Asphalt for laterite and lateritic soils is primarily used for dust control and waterproofing. The more alkaline the soil the less effective the asphalt. To determine the optimum asphalt content and the optimum moisture content for a soil, the Modified AASHO compaction test should be performed as described for a lime- or cement-stabilized soil. A good first approximation of the amount of asphalt to be added to a soil is given herein before in Chapter X. A good rule of thumb is to add about 4 to 6% (by volume) of the void ratio of a soil, or 1/2 gal/yd^2 per in. of the compacted layer. Usually, rapid cure (RC70-800) or medium cure (MC70-800) is applied. The rapid curing asphalt is used to stabilize cohesionless soils and the medium curing asphalt is used to stabilize well-graded soils with some fines. When stabilizing with an asphalt, first scarify the top 6–8 in. of the soil, moisten, apply the asphalt, and then mix. The material is then lightly rolled with a 10-ton vibratory roller, subjected to one pass with a pneumatic roller, and finally finished with a 5- to 8-ton roller. Eight hours later the roadway can be opened to traffic.

Asphalt additives such as annaline and fural with ferric chloride as a catalyst or penlathloropheno significantly increase the strength of an asphalt-stabilized material. These additives are delivered as a slurry and react best in an acid environment. Road tar and lime with an organic acid as a catalyst is another good soil stabilizer combination.

Typical Type of Failures

Rutting, a local shear failure under a wheel loading, is the most common type of failure for a lateritic road or landing strip. The material soil is pushed up and to the side by the wheel. Typically, this condition results from insufficient compaction, and/or water intrusion into the roadway structure either by pumping or percolation.

Local ponding of water is another typical roadway problem. Ponding is caused by poor shaping, concentrated static loadings, differential settlements under traffic, and a fluctuating groundwater table. If ponding is not corrected immediately, traffic will affect serious progressive failures and extensive repair will be required in these areas.

Sloughing, the loss of lateral support for a roadway by an embankment, is another common failure; inadequate compaction and saturation are usual causes.

A roadway may become corrugated when coarse-grained laterites or lateritic soils are used. Differences in density or hardness of a roadway, constant braking and acceleration, and the presence of random rock in the road substructure are the primary causes of corrugation.

The key to preventing these types of failures is to insure adequate and uniform compaction and sufficient drainage facilities to prevent intrusion of water into the roadway structure.

General Suitability for Construction

Laterite

Laterite is suitable as a base, subbase, and select course for a medium-traffic road or airfield (up to C-133) in a forward area (Table 7). Laterite has been used successfully as a 12 in. thick combination base- and surface-course runway for a medium-lift airfield in South Vietnam. Laterite can be used as a surface course and is particularly effective when peneprime or asphalt is applied as a dust palliative and waterproofer. It is excellent as a fill material.

The relatively thin and shallow laterite layer is able to support only light foundation loads. Higher structural loads will require caisson footings which provide good end bearing and require little skin friction. Pile driving into a laterite is very difficult and requires steel shoes on the piling. Penetration is usually no more than a few feet and pile load tests should be performed in all cases to check design requirement compliance. Laterite is not a substitute for good rock. When laterite is used as a concrete aggregate (which is not recommended because of its tensile strength) flexure tests must be performed to insure that the design strengths are achieved. Blocks cut from laterite can be used effectively as building material.

Lateritic Soils

Most lateritic soils make good subgrades and can be used as a subbase material if certain precautions are taken. These soils are good for fill material and compacted fill in earth dams used for impounding water.

Table 7. Design Criteria for Roads and Airfields

Material	Maximum design, CBR	Maximum particle size, in.	Gradation required in % passing		Liquid limit	Plasticity index
			#10	#200		
Base course	50	—	—	10	25	5
Subbase	50	2	50	15	25	5
Subbase	40	2	80	15	25	5
Subbase	30	2	100	15	25	5
Select material	20	3	—	—	35	12

Use of Laterites and Lateritic Soils for Construction "Rules of Thumb"

The recognition of outcrops or hardened massive crystallized laterite, vascular laterite, or pelletized laterite should present little or no difficulty in the field to the construction engineer. These materials can be used readily in all phases of road and airfield construction under dry base and surface courses. Stabilization of surface courses can be readily accomplished by bituminous penetration if time, equipment, and material permits. Laterites also may be used as aggregates for concrete mix, if rigid pavement is desired; however, the flexural strength may be low.

The recognition and use of lateritic soils present a special problem to the engineer in the field, particularly in the case where little or no experience is involved. Lateritic soils may run the gamut of soil classification from SP to CH.

In the field, under certain conditions, the construction engineer may find normal means of soil classification unavailable. Orders are to commence construction of a road (or airfield) immediately utilizing such equipment and construction materials available on or near the site of operations. There is no soils analysis equipment available and the construction is to be accomplished by a newly arrived unit unfamiliar with the work area and its surrounding topography. What to do? Let's examine some "rules of thumb" which may resolve the immediate problem.

a. Recon the job first. Choose the point of start for clearing vegetation and overburden.

b. Commence the clearing of vegetation and overburden while selecting the borrow site for construction material.

c. Prepare a drainage plan. Culvert to natural existing drainage gulleys or channels where practical. Do not, if at all possible, interrupt the natural drainage pattern. Place your drainage to conform to this requirement. Drainage culverts, ditches, etc. should be built concurrent with other construction progress.

d. The borrow site(s) already have been selected. In the case where good borrow material is not available, construction has to be accomplished with such soil material available. This borrow material may be of questionable classification, especially if it is described as a lateritic soil. In the absence of soils analysis equipment, choice of borrow material may be based on:

1. The ball test—drop a ball or dry lump of the soil into a container of water. If it disintegrates rapidly, it is not suitable for use.
2. Soil hardness—is the dry lump hard by feel, does it have roughness with sharp edges or does it feel soft, smooth, and powdery? If it feels soft, smooth, and powdery, do not use this material.
3. Weight—is the dry soil lump heavy by feel, does it appear to be matted with hard, reddish-brown modules or is it light to the feel and just reddish in color? If it feels lighter in comparison with other borrow material available, do not use this material.
4. Grittiness—does the dry lump feel gritty or coarse? If so, sands and small gravelly materials are present and material is usable.

e. Optimum moisture and compaction. Spread your borrow material on your construction site and begin compaction. If the field moisture is low, add water gradually while compacting, testing frequently by balling compacted soil in your hands until a firmness point is reached where slight finger imprint is left on the ball. At this point the soil is near its optimum moisture content. You have also established the water to dry borrow material ratio required to continue construction. Continue compaction until the surface appears even, smooth, and hard. (With lateritic soils, light compaction is recommended. Heavy compaction destroys the drainage characteristics of these soils, making them highly plastic with the resulting loss of bearing strength.)

Drive a twenty-penney nail into the compacted surface at several random points. If the nail "pings" to the hammer blows, compaction is good. Caution —be sure and drive the nail all the way down to the compacted surface to be sure of underlayer compaction.

f. Untreated surface—road(s) or airfield(s). Crown the surface—use 4% slope to ensure adequate runoff during rain. Palliate, if possible, with salt or oil. Bituminous or asphalt surfacing will depend on duration of intended usage and will be a higher command decision.

REFERENCES

1. Buchanan, F. A. *A Journey from Malabar through the Countries of Mysore, Canana and Malabar*, London, East India Company, 12p (436-460), 180F.
2. Carvalho, G. S. The genesis of 'laterites' from the geological point of view, *Proceedings of the 4th Regional Conference for Africa on Soil Mechanics and Foundation Engineering*, Cape Town, 1967.
3. de Medina, Jacques. *Laterites and their Application to Highway Construction*, based on the text of a course at the Brazilian Highway Research Institute.
4. Persons, Benjamin S. A discussion of "the genesis of 'laterites' from the geological point of view" by G. S. Carvalho, *Proceedings of the 4th Regional Conference for Africa on Soil Mechanics and Foundation Engineering*, Cape Town, 1967.
5. Voloboyev, V. R. *TICSS*, 6, Vol. 20, 1956; *Sov. Soil. Science* No. 11, 1961.
6. *UNESCO Review of Research on Laterite.*
7. Florentin, J., and L'Heriteau G. A study of the physio-chemical characteristics of laterite, Vol. I, *Proceedings of the Fourth International Congress of Soil Mechanics*, London, 1957.
8. D'Hoore, J. *Studies in the Accumulation of Sesquioxides in Tropical Soils*, National Institute for Agronomy in the Belgian Congo, Scientific Series No. 62, 1954.
9. Pendleton, R. L., and Shanasuvana, S. Analyses of some Siamese laterites, *Soil Science*, pp. 423–440, 1946.
10. Cofer, H. E.,Jr. *Laterites in the Georgia Coastal Plain*, Consulting Report, 1967.
11. Liang, Ta. *Tropical Soils. Characteristics and Airphoto Interpretation* Unclassified Report for the United States Air Force, AFCRF 64–937, dated August 31, 1964.
12. Lukens, J. E. *Chemical Properties of Tropical Soils*, Cornell University Unpublished Report, 1964.
13. Caiger, J. H. The use of airphoto interpretation in materials investigation for rural road projects, *Proceedings of the 4th Regional Conference for Africa on Soil Mechanics and Foundation Engineering*, Cape Town, 1967.
14. Brink, B. A. *Airphoto Interpretation Applied to Soil Engineering Mapping in South Africa*, International Symposium on Photo Interpretation, Delft, 1962.
15. Holden, A. Laterites: identification and interpretation by airphotos, *Proceedings of the 4th Regional Conference for Africa on Soil Mechanics and Foundation Engineering*, Cape Town, 1967.
16. Loureiro, F. E. V. L. Some results obtained in Angola in the survey of soils by photo interpretation, *Proceedings of the 4th Regional Conference for Africa on Soil Mechanics and Foundation Engineering*, Cape Town, 1967.
17. Mountain, M. J. The location of pedogenic materials using aerial photographs, with some examples from South Africa, *Proceedings of the 4th Regional Conference for Africa on Soil Mechanics and Foundation Engineering*, Cape Town, 1967.
18. *Cornell University Navy Report, Land Form Series*, 1951.
19. Cheney, W. O., and Richards, Gordon V. Ocean temperature measurement for power plant design, *Coastal Engineering*, 1966.

INDEX